LEONARDVS EVLER.
qui
cognitione sua naturæ arcana reclusit.

J. a Si.

Brucker pinx! Petropoli 1737.

# Leonhard Euler

**Emil A. Fellmann**
Translated by Erika Gautschi and Walter Gautschi

Birkhäuser Verlag
Basel · Boston · Berlin

Author:

Emil A. Fellmann
Arnold Böcklin-Strasse 37
CH-4051 Basel
Switzerland

Library of Congress Control Number: 2006937470

Bibliographic information published by Die Deutsche Bibliothek
Die Deutsche Bibliothek lists this publication in the Deutsche Nationalbibliografie; detailed bibliographic data
is available in the Internet at <http://dnb.ddb.de>.

ISBN 978-3-7643-7538-6 Birkhäuser Verlag, Basel – Boston – Berlin

© 2007 Birkhäuser Verlag, P.O. Box 133, CH-4010 Basel, Switzerland
Part of Springer Science+Business Media
Printed on acid-free paper produced from chlorine-free pulp. TCF ∞
Printed in Germany
ISBN-10: 3-7643-7538-8
ISBN-13: 978-3-7643-7538-6                                        e-ISBN: 978-3-7643-7539-3

9 8 7 6 5 4 3 2 1                                                www.birkhauser.ch

In memoriam
Christian Fellmann
1958–1987

# Contents

The solar system in the cosmos. Engraving by Berot after F. K. Frisch in the frontispiece to Euler's "Theory of the planets and comets", Berlin 1744. Note the trajectory of the comet (Halley), indicated by a dotted path passing between the sun and Mercury, on which the tail is correctly pointing away from the sun. The earth with its satellite can be seen vertically below the sun, Jupiter and Saturn with their four (known at the time) moons each to the right resp. left of the central star.

# Preliminary remarks

A book about a great mathematician which is entirely free of formulae may appear to specialists perhaps a little strange at first. They should consider, however, that—according to the intentions of the original publisher—the present work has been written for a broad audience with interests in the history of culture and science. The historically inclined mathematician, physicist, or astronomer will easily find access to EULER's technical discoveries with the help of the many sources provided in notes and in the bibliography, which enable a deeper penetration into the material; another easy access is provided, for example, by the synoptically laid out "Basler Euler-Gedenkband" of 1983 (EGB 83). The four subsections identified with a * presuppose a certain amount of technical knowledge and, if necessary, may be skipped by the reader without loss of continuity.

LEONHARD EULER's unusually rich life and broadly diversified activity in the immediate vicinity of important personalities which, in the truest sense of the word, have made history, may well justify an exposition, like the present one, which in part is based on unpublished sources and comes, as it were, right out of the workshop of the current research on EULER.

Two technical details must be pointed out. In correspondence, there occasionally are double dates. The reason for this is the fact that in Russia, until February of 1918, the (old) Julian calendar was still in force, in Prussia and elsewhere, however, the (new) Gregorian calendar. In the 18th century, the difference amounted to eleven days by which the Gregorian calendar "ran ahead" of the Julian. Accordingly, in double dates, the first indicates the day in the old style, the second the same day in the new style. Finally, avoiding the Cyrillic alphabet requires transliteration of Russian names and work titles.

# Prologue

Elephants are drawn always smaller
than life, but a flea always larger.
JONATHAN SWIFT

Basel, St. Petersburg, and Berlin determine exactly the three points of the historical plane in which LEONHARD EULER's life transpired, and the name of his place of birth, Basel, is well known, even famous, in the history not only of humanism, but also of the mathematical sciences: The brothers JAKOB and JOHANN BERNOULLI, in the twilight of the 17th century, illuminated the mathematical horizon as a double-star of the greatest order, the mightily driving seeds of the LEIBNIZ infinitesimal calculus — a family secret, as it were — enabling them to largely monopolize mathematics well into the next century. Their brilliance was to be outshined only by the "sun of all mathematicians of the 18th century", as EULER was called, who as a phenomenon stood equally unparalleled in the history of science as his native town in the history of Europe.

One does Basel an injustice with the often expressed reproach of having treated EULER unduly badly and not having recognized his genius early enough, merely by not having made the not yet twenty-year-old young man a professor right away when, early in 1727, he applied for the chair in physics. Because, in the first place, except for his 16-page "Habilitationsschrift" on the theory of sound, he had published only two small papers of three resp. five printed pages, and secondly, his teacher JOHANN BERNOULLI was the only one who was able to recognize the unusual talents which were lying dormant in the young LEONHARD. When BERNOULLI, who after ISAAC NEWTON's death (1727) moved up to become the first mathematician of the world, died in 1748 at the age of 81, EULER indeed was immediately called to Basel — the rather complicated election procedure having been bypassed — but he declined with thanks: In the meantime, he had found an arena of activity in the big world which was commensurate with his thirst for action and his genius and, in a manner of speaking, he personified the two "mammoth-academies" of Berlin and Petersburg. In the 18th century the city of Basel commissioned the sculptor HEINRICH RUF (1785) to make a marble bust of LEONHARD EULER, which today is located in the entrance hall of the Bernoullianum, and in 1871 named a street after its great son, which — only by accident, I am sorry to say — is the continuation of the LEONHARD-Strasse; thus Basel came to have a "LEONHARD-EULER-Strasse". But as nicest tribute paid to LEONHARD EULER by the city of Basel one has to mention the handsome memorial volume[1] published on the 200th anniversary of EULER's death; on 555 pages, it collects thirty contributions by 28 scholars from ten nations and four continents — all works which, according to newest research on EULER's life, cover the unusually broad spectrum of EULER's scientific activities in a well thought-out plan.

On the helvetic level, the greatest citizen of Basel was remembered in 1979 with the smallest banknote of widest circulation, and the publication of EULER's collected works[2], a multimillion enterprise, since 1907 is generously supported and promoted not only by private industry and learned societies of several nations, but also by Swiss federal agencies; we have in mind, in this connection, the Swiss Academy of Natural Sciences and also the Swiss National Science Foundation.

EULER was not only by far the most productive mathematician in the history of mankind, but also one of the greatest scholars of all time. Cosmopolitan in the truest sense of the word — he lived during his first twenty years in Basel, was active altogether for more than thirty years in Petersburg and for a quarter of a century in Berlin — he attained, like only a few scholars, a degree of popularity and fame which may well be compared with that of GALILEI, NEWTON, or EINSTEIN. For this reason, the demand on a biographer to not let a biography degenerate into a hagiography is, in the case of EULER, especially difficult to comply with.

With regard to EULER's character, all contemporaries and biographers are unanimous: He was a child of the sun, as astrologers would say, with an open and cheerful mind, uncomplicated, humorous and sociable. Even though wealthy to rich in the second half of his life, he was modest in material affairs, always free of any conceit, never vindictive, but self-assured, critical, and daring. At times he could flare up a bit, only to calm down immediately, even to laugh about his own outburst. In one point, however, he wouldn't stand for any nonsense: in the matter of religion and Christian faith. EULER's orthodoxy is the key to understanding many important facts in his life, for example his relentless persecution of LEIBNIZ's doctrine of monads in the vein of WOLFF, as also his severe attacks against certain encyclopedists and other "free-thinkers", which he launched 1747 in his theological pamphlet *Rettung der göttlichen Offenbarung* [Salvation of the divine revelation]. Nevertheless, EULER's (practiced) tolerance was by far more honest and prominent than the one of his royal master FREDER-ICK II, who used it only as catchword and propaganda word, and could forget it on the spot when practicing it would have been in the slightest inconvenient.

Also matters of scientific priority were foreign to EULER: Contrary to most scholars of any time, he never knew priority quarrels; indeed, at times he generously gave away new discoveries and insights. In his works he doesn't hide anything, but lays the cards always open on the table and offers the readers the same opportunities and chances of finding something new; indeed he often leads them very close to the discovery and surrenders the joys of discovery to them — the only true pedagogy. This makes EULER's books an adventure for the student, entertaining and exciting at the same time. The feeling of envy must have been absolutely foreign to this astonishing human being; he granted everything to everyone and was always delighted also at the new discoveries of others. This all was possible for him only because he was spiritually so immensely rich and psychologically well-balanced to a degree rarely found.

The phenomenon EULER is essentially tied to three factors: first to the gift of a possibly unique memory. Anything EULER ever heard, saw, or wrote, seems to have been firmly imprinted forever in his mind. For this, there are numerous contemporary testimonials. Still at an advanced age, he was known, for example, to delight members of his family, friends and social gatherings, with the literal (Latin) recitation of any song whatsoever from VERGIL's *Aeneis*, and protocols of the Academy meetings he still knew by heart tens of years later — not to speak of his memory for matters in mathematics. Secondly, his enormous mnemonic power was paired with a rare ability of concentration. Noise and hustle in his immediate vicinity barely disturbed him in his mental work: "A child on his knees, a cat on his back, this is the way he wrote his immortal works" reports his colleague THIÉBAULT. The third factor of the "mystery EULER" is simply steady, quiet work.

# 1
## Basel 1707–1727

## Leonhard Euler's autobiography

How could one introduce the biography of a famous person in a more appropriate manner if not with a brief self-testimonial, if one exists? For Leonhard Euler, this is fortunately the case, since soon after the beginning of the second Petersburg period, Euler — presumably at the urging of members of the family — dictated to his son, Johann Albrecht, a short curriculum vitae in the German language[3], which we here reproduce in the original orthography [then in free English translation]:

> "Meines Vaters Lebens-Lauf so wie er ihn mir selber in
> die Feder dictirt hatte.
> Geschrieben zu St. Petersburg den 1$^n$ December 1767."
>    *Ich, Leonhard Euler, bin A. 1707 den 15$^{ten}$ April*
> *St. nov. [new style] zu Basel gebohren. Mein Vater war Paulus*
> *Euler, damahls designirter Prediger nach dem*
> *eine Stund von Basel gelegenen Dorf Riechen: und*
> *meine Mutter hiess Margaretha Bruckerin. Bald*
> *hierauf begaben sich meine Eltern nach Riechen,*
> *wo ich beÿ Zeiten von meinem Vater den ersten*
> *Unterricht erhielt; und weil derselbe einer von den*
> *Discipeln des weltberühmten Jacobi Bernoulli ge-*
> *wesen, so trachtete er mir sogleich die erste*
> *Gründe der Mathematic beizubringen, und*
> *bediente sich zu diesem End des Christophs Rudolphs*
> *Coss mit Michaels Stiefels Anmerckungen, wo-*
> *rinnen ich mich einige Jahr mit allem Fleiss übte.*
> *Beÿ zunehmenden Jahren wurde ich in Basel*
> *beÿ meiner Grossmutter an die Kost gegeben,*
> *um theils in dem Gymnasio daselbst, theils durch*

*Privat Unterricht den Grund in den Humanioribus
zu legen und zugleich in der Mathematic weiter
zu kommen. A. 1720 wurde ich bey der Universität
zu den Lectionibus publicis promovirt: wo ich
bald Gelegenheit fand dem berühmten Professori
JOHANNI BERNOULLI bekannt zu werden, welcher sich
ein besonders Vergnügen daraus machte, mir
in den mathematischen Wissenschafften weiter
fortzuhelffen. Privat Lectionen schlug er mir zwar wegen
seiner Geschäfte gänzlich ab: er gab mir aber einen weit
heilsameren Rath, welcher darin bestand, dass ich selb-
sten einige schwerere mathematische Bücher vor mich
nähmen, und mit allem Fleiss durchgehen sollte, und
wo ich einigen Anstoss oder Schwierigkeiten finden
möchte, gab er mir alle Sonnabend Nachmittag einen
freyen Zutritt bey sich, und hatte die Güte mir die
gesammlete Schwierigkeiten zu erläutern, welches
mit so erwünschtem Vortheile geschahe, dass wann er
mir einen Anstoss gehoben hatte, dadurch zehn andere
auf einmahl verschwanden, welches gewiss die beste
Methode ist, um in den mathematischen Wissenschafften
glückliche Progressen zu machen.
   A. 1723 wurde ich zum Magister promovirt, nachdem
ich anderthalb Jahr vorher nach der dasigen Gewohnheit
primam lauream erhalten hatte. Nachdem musste ich
mich auf Gutbefinden meiner famille beÿ der
Theologischen Fakultaet einschreiben lassen, da ich
mich denn ausser der Theologie besonders auf die
griechische und Hebräische Sprache appliciren sollte:
womit es aber nicht recht fort wollte, weil ich meine
meiste Zeit auf die mathematische Studia wendete,
und zu meinem Glück die Sonnabend Visiten
bey dem Herrn JOHANNI BERNOULLI noch immer
fortdaureten. Um dieselbige Zeit wurde die
neue Academie der Wissenschafften in St. Petersburg errichtet, wohin die beyden ältesten*

*Söhne des H. JOHANNIS BERNOULLI beruffen wurden;*
*da ich denn eine unbeschreibliche Begierde bekam*
*mit denselben zugleich A. 1725 nach Petersburg*
*zu reisen. Die Sache konnte aber damahls nicht so*
*gleich zu Stande gebracht werden. Die gemeldten*
*jüngern BERNOULLI gaben mir indessen die feste Versicherung,*
*dass sie mir nach ihrer Ankunft in Petersburg eine*
*anständige Stelle daselbst auswürcken wollten, welches*
*auch würcklich bald darauf erfolgt, da ich um meine*
*mathematische Kenntnüss auf die Medicin zu appliciren*
*bestimmt wurde. Weil diese Nachricht zu Anfang des*
*Winters A. 1726 einlief, und ich meine Abreise nicht*
*vor dem künftigen Frühjahr vornehmen konnte, so liess*
*ich mich inzwischen bey der medicinischen Fakultaet in Basel*
*immatriculiren und fing an mich mit allem Fleiße*
*auf das Studium medicum zu appliciren: inzwischen*
*wurde zu Basel die Professio Physica erlediget,*
*und weil sich dafür eine Menge Competenten meldete,*
*so liess ich mich auch in die Zahl derselben aufschreiben,*
*und hielte bey dieser Gelegenheit als Praeses meine*
*Disputationem de Sono. Inzwischen rückte das*
*Frühjahr A. 1727 heran, und ich trat meine Abreise*
*von Basel gleich im Anfang des April Monaths an,*
*kam auch so früh nach Lubec, dass noch kein Schif*
*fertig lag um nach Petersburg zu segeln: ich*
*war allso gezwungen mich auf ein nach Reval*
*gehendes Schif zu setzen, und weil die Reise*
*gegen vier Wochen daurete, so fand ich in Reval*
*bald ein Stettiner Schif, so mich nach Cronstadt*
*transportirte. Daselbst aber kam ich an eben*
*demjenigen Tage an, da die hochsel: Kayserin CATHERINA*
*1. ALEXIEVNA Todes verblichen war, und fand allso*
*in Petersburg bey der Academie alles in der*
*grössten Consternation. Doch hatte ich das Vergnügen*
*ausser den jüngeren H. DANIEL BERNOULLI, indem*
*sein älterer Herr Bruder NICOLAUS inzwischen ver-*

3

storben war, noch den sel. H. Prof. HERMANN, welcher ebenfalls mein Landsmann, und noch dazu ein
weitläuffiger Verwandter von mir war, anzutreffen,
welche mir allen nur ersinnlichen Vorschub thaten. Meine Besoldung war 300 Rbl. nebst freyer Wohnung, Holtz und Licht,
und da meine Neigung einig und allein auf die mathematischen Studien gerichtet war, so wurde ich zum Adjuncto
Matheseos sublimioris bestellt, und der Vorschlag mich
beÿ der Medicin zu employiren fiel gänzlich weg.
Wobeÿ mir die Freyheit erteilt wurde den academischen
Versammlungen mit beÿzuwohnen, und daselbst meine
Ausarbeitungen vorzulesen, welche auch schon damahls
den academischen Commentarien einverleibet wurden.
Als hierauf A. 1730 die Herren Professores HERMANN
und BÜLFINGER wieder in ihr Vaterland zurück kehrten,
so wurde ich an des letzteren Stelle zum Professore
Physices ernannt, und machte einen neuen Contract
auf 4 Jahre, nach welchem mir die zweÿ ersteren Jahren
400 Rbl., die zweÿ letztere Jahre aber 600 Rbl. nebst 60 Rbl.
für Wohnung, Holtz und Licht accordirt wurden. Zur Zeit
dieses Contracts verheÿrathete ich mich A. 1733 um Weÿnachten
mit meiner frau CATHERINA GSELL, und da um dieselbe
Zeit der Hr. Prof. DANIEL BERNOULLI auch nach seinem
Vaterlande zurückgereiset, so wurde mir seine Professio
Matheseos sublimioris aufgetragen, und bald darauf
erhielt ich von dem dirigirenden Senat den Befehl auch die
Aufsicht über das geographische Departement zu übernehmen,
beÿ welcher Gelegenheit mir meine Besoldung auf 1200 Rbl.
vermehrt wurde. Als hierauf A. 1740 Seine noch glorreich
regierende Königl: Majestät in Preussen zur Regierung kamen,
so erhielt ich eine allergnädigste Vocation nach Berlin, welche ich auch,
nachdem die glorwürdige Kayserin ANNE verstorben war, und
es beÿ der darauf folgenden Regentschafft ziemlich misslich
auszusehen anfieng, ohne einiges Bedencken annahm, und nach
erhaltenem Abschied A. 1741 mich mit meiner ganzen famille
zu Wasser nach Berlin verfügte, wo Se. Königl: Majestät

*mir eine Besoldung von 1600 Thl. als ein Aequivalent der hier*
*genossenen Gage festzusetzen geruhte. Was mir darauf*
*weiter begegnet, ist bekannt.*

["My father's life as dictated to me by him. Recorded in St. Petersburg on the 1st of December 1767".

I, LEONHARD EULER, was born in Basel on the 15th of April (new calendar) 1707. My father was PAULUS EULER, then designated minister in the village of Riehen, an hour away from Basel, and my mother's name was MARGARETHA BRUCKER. Soon thereafter my parents moved to Riehen where in due time I received from my father my first tuition; and because he had been one of the disciples of the world-renowned JACOB BERNOULLI, he tried to impart to me the first principles of mathematics, and to this end used CHRISTOPH RUDOLPH's *Coss*, with annotations by MICHAEL STIEFEL, from which I practiced diligently for several years. In subsequent years I boarded with my grandmother in Basel in order to learn the basics in the humanities, partly at the local "Gymnasium", partly through private tuition, and at the same time to make progress in mathematics. In 1720 I was admitted to the university as a public student, where I soon found the opportunity to become acquainted with the famous professor JOHANN BERNOULLI, who made it a special pleasure for himself to help me along in the mathematical sciences. Private lessons, however, he categorically ruled out because of his busy schedule: However, he gave me a far more beneficial advice, which consisted in myself taking a look at some of the more difficult mathematical books and work through them with great diligence, and should I encounter some objections or difficulties, he offered me free access to him every Saturday afternoon, and he was gracious enough to comment on the collected difficulties, which was done with such a desired advantage that, when he resolved one of my objections, ten others at once disappeared, which certainly is the best method of making auspicious progress in the mathematical sciences.

In 1723 I was promoted to a *magister* after I had received, one-and-a-half years earlier, according to prevailing customs, the *primam lauream*. Thereafter, at the discretion of my family, I had to register at the Theological Faculty, since I was then expected to apply myself not only to theology but especially also to the Greek and Hebrew languages, which however did not get on very well since I devoted most of my time to my mathematical studies and, fortunately, the Saturday visits with JOHANN BERNOULLI still continued. At the same time, the new St. Petersburg Academy of Sciences was founded to which the two eldest sons of JOHANN BERNOULLI were called; upon which I became indescribably eager to travel with them both, already in 1725, to Petersburg. The matter, however, at the time could not be materialized so quickly. The two younger BERNOULLI, nevertheless, made the firm promise that, after their arrival in Petersburg, they would procure for me a decent position there, which indeed happened soon thereafter, as I was designated to apply my mathematical knowledge to medicine. Since this news came to me at the beginning of the winter of 1726 and I could not arrange my departure before the following spring, I enrolled at the Medical Faculty in Basel and began to apply myself with all my diligence to the study of medicine: In the meantime, the professorship of physics became vacant in Basel, and since many candidates applied for this position, I let my name also be included among theirs and on this occasion presented as *Praeses* my *Disputationem de Sono*. In the meantime, the spring of 1727 drew near and I set out from Basel right at the beginning of April, but arrived in Lubec at such an early time that no ship was ready to sail to Petersburg: I was forced therefore to take a ship going to Reval, and since the trip took almost four weeks, I soon found in Reval a ship to Stettin which brought me to Cronstadt. There, I arrived on precisely the day when the death of the Empress CATHERINA I ALEKSEYEVNA became known and thus found in Petersburg at the Academy everything in greatest consternation. Yet I had the pleasure to meet, besides the younger DANIEL BERNOULLI, his older brother NICOLAUS having passed away in the meantime, still the late Prof. HERMANN, also a compatriot and, besides, a distant relative of mine, who gave me all imaginable assistance. My salary was 300 Rbl. along with free lodging, firewood, and light, and since my inclinations were directed solely and exclusively toward mathematical studies, I was appointed as an adjunct of higher mathematics, and the suggestion to occupy myself with medicine was dropped altogether. At the same time I was given the liberty of attending the sessions of the Academy and to present there the results of my work, which already at that time were incorporated in the *Commentaries* of the Academy. When thereafter, in 1730, the Professors HERMANN and BÜLFINGER returned to their native country,

I was appointed in the latter's place as a professor of physics, and entered in a new contract for 4 years, according to which I was awarded for the first two years 400 Rbl., but for the last two years 600 Rbl. along with 60 Rbl. for lodging, firewood, and light. During the course of this contract, I married Catherina Gsell in 1733 around Christmas, and since at this time also Prof. Daniel Bernoulli returned to his native country, his Chair of mathematics was entrusted to me, and soon thereafter I received from the senate the order to also take over the supervision of the Department of Geography, at which occasion my salary was increased to 1,200 Rbl. Then, in 1740, when His still gloriously reigning Royal Majesty came to power in Prussia, I received a most gracious call to Berlin, which, after the illustrious Empress Anne had died and it began to look rather dismal in the regency that followed, I accepted without much hesitation, and after having received my discharge, moved with my entire family on water to Berlin, where His Royal Majesty was pleased to fix my salary at 1,600 Thl., the equivalent of the honorarium which I had the fortune to receive here. What happened to me later is known.]

## Euler's genealogy

A thorough and exemplary presentation of the genealogy of Euler's entire family is in hand since 1955, written by the theologian Karl Euler (1877–1960)[4]. He traced the origin of Euler's family back to the 13th century — ignoring, to be sure, the feminine branches. The family name Euler (Öwler) is first mentioned 1287 in Lindau at Lake Constance, has been securely ascertained, however, since 1458. The family carried a double-name of variable spelling, today most frequently written Euler-Schölpi. The second name is derived from the Alemannic "schelb", which means "crooked", "oblique", but also "cross-eyed", and figuratively also "small rascal"; it was discontinued only by Hans-Georg Euler (1573–1663), the great-grandfather of Leonhard Euler, as he settled in Basel as combmaker and acquired there, in 1594, the civil rights. The surviving root name Euler is derived from "Ouwe" (Aue, Au, a small swampy meadow), and an "Ouwler", "Owler" in the Alemannic area was an owner of such. Probably

PROSPECT DER RHEINBRÜCKE ZU BASEL, VUE DU PONT DU RHIN DE BASLE
VON SEITEN DER KLEINEN STADT. DU CÔTÉ DE LA PETITE VILLE.

Em. Büchel del. 1761. (D. Herrliberger exc. Can. Priv.

Basel in the year 1761. Engraving by W. HERRLIBERGER after a drawing by E. BÜCHEL

quite erroneous is the derivation of the name EULER from the Roman times in connection with the craft of potter (Lat. olla, pot, Middle High German "aul", clay, potting soil).

Recently, in a Soviet omnibus volume[5], there appeared a kind of continuation of the work of KARL EULER. The three authors of this "new genealogy" made it their task to substantially improve and complete the annotations in the earlier work, and — as far as it is possible today — to list all descendants of LEONHARD EULER, also in the feminine branches which do not carry the name EULER. The authors succeeded in eliciting more than a thousand descendants of LEONHARD EULER, who all carry resp. carried his name; of these, about 400 are still alive today, more than half of them in Russia and — according to verbal information from the former parliamentarian ALEXANDER EULER in Basel — 16 resp. 29 in Switzerland.

# The parents

Unfortunately we don't know much more about LEONHARD EULER's mother than what MICHAEL RAITH assembled in his competently written article about the father PAULUS EULER.[6] MARGARETHA BRUCKER (1677–1761) was the daughter of the hospital vicar JOHANN HEINRICH BRUCKER (1636–1702); among her ancestors were an impressive line of men, highly educated in the classics, as for example the Latinist CELIO SECONDO CURIONE (1503–1569), the Hebraist JOHANNES BUXTORF (1564–1629), the lawyer and highest guild master BERNHARD BRAND (1523–1594), and also the famous ZWINGER dynasty of scholars.[7]

PAULUS EULER (1670–1745), LEONHARD's father, could not enjoy such an illustrious gallery of ancestors, and in this regard he was a *homo novus*, which certainly can no longer be said of his son, who in fact "upholds the heritage of the humanist city of Basel" (RAITH). PAULUS EULER's father of the same name (1638–1697) was the grandson of HANS-GEORG EULER mentioned above, himself a combmaker. Although he appears 1654 in the registers of the University of Basel, nothing is known about any academic degree. He married in 1669 ANNA MARIA GASSNER (1642–1712), the daughter of a pastry baker who immigrated from Vöcklabruck in upper Austria.

After the death of her husband, she lived for the rest of her life with her son PAULUS in Riehen.

At the time of EULER's youth, Basel was one of the thirteen republics of which Switzerland was then made up; around 1725 it had about 17,000 inhabitants, among them a good many descendants of Calvinists who had been driven out of France and Italy. In the rather conservative Basel of the 18th century hardly any new citizens were taken in; there was a supremacy of the guilds, the political and economic power was concentrated in a few families, church and state functioned as an undisputed unity, and attendance in public worship was considered almost primarily an expression of good faith in the authorities. But side by side with conventional forms of faith and thought, there was also abrupt critical questioning and skeptical distance.[8]

PAULUS EULER, according to entries in the register, enrolled 1685 at the University, founded in 1460 by Pope PIUS II (PICCOLOMINI) as a sequel to the Council of Basel (1431–1448), and after a propaedeutic *studium generale* at the Philosophical Faculty he eventually chose (protestant) theology as his special field of study. During the first semesters, PAULUS EULER also engaged in mathematical studies, in the course of which he participated in October of 1688 in a debate on *Positiones mathematicae de rationibus et proportionibus* chaired by the great mathematician JACOB BERNOULLI. These fifteen theorems or postulates, though, are not authored by PAULUS EULER, but by JACOB BERNOULLI, and were reprinted in his posthumous works published by the nephew NIKLAUS I BERNOULLI (1744), after they had appeared in print already 1688 in Basel.[9] Following this debate, PAULUS EULER took up the study of theology, earned the academic degree of Masters in 1689, and became 1693 *Sacri Ministerii Candidatus*, which is equivalent to becoming eligible for the ministry. Even though his first benefices — 1701 the incumbency at the penitentiary and orphanage in Basel and from 1703 the one of St. Jakob at the Birs — were paid rather poorly, he could marry MARGARETHA BRUCKER in 1706. From her he had four children, of whom LEONHARD was the first-born. There were two sisters that followed: ANNA MARIA (1708–1778) and MARIA MAGDALENA (1711–1799), then the brother JOHANN HEINRICH (1719–1750)[10], who as a student of GEORG GSELL (1673–1740), the future father-in-law of LEONHARD EULER in St. Petersburg, was to become an artist. The external conditions of PAULUS EULER improved after his appointment on June 27, 1708 to the ministry in Riehen, where in November he moved into the rather cramped[11] residential premises assigned to him in the "Leutpriester" house, which still stands today, although structurally remodeled in 1851. There, he should remain until his death on March 11, 1745, in faithful and conscientious compliance with his manifold duties, esteemed and loved by his parish.

# Childhood and youth

LEONHARD EULER was born on Friday, April 15, 1707 — most probably in the center of Basel, certainly not in Riehen. His birthplace is not known, since his father then was still a vicar at St. Jakob. This picturesque group of houses, though lying outside the old city walls, was still inside the limits of the municipality of Basel and had a small church — still standing today — but no rectory. At the time of LEONHARD's birth, the EULER family presumably lived in the vicinity of the church of St. Martin, where indeed the baptism took place on April 17, 1707. LEONHARD — so named after RESPINGER, one of the three godfathers — was about one-and-a-half years old when the family moved into the rectory in Riehen.

Riehen is situated in a pocket of land north of the Rhine river belonging to Basel but surrounded by German territory, about halfway between Basel and Lörrach — an hour's march "on the most boring street of the world", which LEONHARD as a schoolboy surely must have trodded along many times. This charming village, still today famous for its cherries and vineyards, may then have numbered about 1,000 inhabitants[13] and was the rural, quiet scene of LEONHARD's childhood. Not much is known about the latter, but among the little we know, there is a charming, yet authentic, little story of a contemporary about LEONHARD EULER, which we relate here in free translation from the Latin: "As a small boy about four years old, living in the country, he [LEONHARD] observed how the hens, sitting on eggs, hatched their chickens and in this way brought forth their young to light. In the hope of accomplishing something similar, he secretly collected the eggs from the nests, put them down in a corner of the chicken coop, sat on them, and didn't let go until — missed for hours, and anxiously sought by his parents — he was eventually found over the eggs ... When asked what

JAKOB I BERNOULLI. Oil painting by NIKLAUS BERNOULLI, 1694

JOHANN I BERNOULLI. Engraving by J. J. HAID after an oil painting by J. R. HUBER, around 1740, detail

in the world he was doing here, the little chap replied: 'I want to make young chickens'."[14]

PAULUS EULER gave his son the first elementary instruction, and LEONHARD's first mathematical textbook was the *Coss*, MICHAEL STIFEL's edition (1553) of CHRISTOFF RUDOLFF's *Algebra* of 1525[15] — an exceedingly difficult book for a boy of LEONHARD's age.

Presumably in the eighth year of his life, LEONHARD was sent to the Latin school in Basel, where he was put out to board and lodge at his widowed, maternal grandmother MARIA MAGDALENA BRUCKER-FABER. The "Gymnasium" in Basel was then in a rather dismal condition[16], since other than Latin and (optionally) Greek, hardly anything else could be learned there. For example, mathematics as a subject of study was cut upon the request of the citizenry, in spite of many vehement appeals of the world-renowned mathematician JOHANN BERNOULLI (1667–1748), in his capacity as inspector of the educational system, to make improvements. Like many parents who worried about an efficient education and further development of their children, PAULUS EULER therefore engaged for his son a private tutor, namely the young theologian JOHANNES BURCKHARDT (1691–1743), a future vicar in Kleinhüningen, then in Oltingen in the countryside around Basel. He, too, was an enthusiastic mathematician, and his influence on the young LEONHARD EULER — though not yet clarified in detail — must have been very significant, since DANIEL BERNOULLI (1700–1782), the congenial son of JOHANN and future friend of LEONHARD EULER, characterized BURCKHARDT in a letter containing the announcement of his death as "teacher in mathematics of the great EULER".

## The time in Basel until 1727

At the age of thirteen, an age quite normal in the circumstances of that time, LEONHARD, who according to the notion of his parents was to become a theologian, enrolled at the University of Basel. At its lower level, the university then had also propaedeutic functions, which nowadays belong in the province of the upper-level "Gymnasium". In October of 1720, LEONHARD EULER enrolled, for the time being, at the Philosophical Fac-

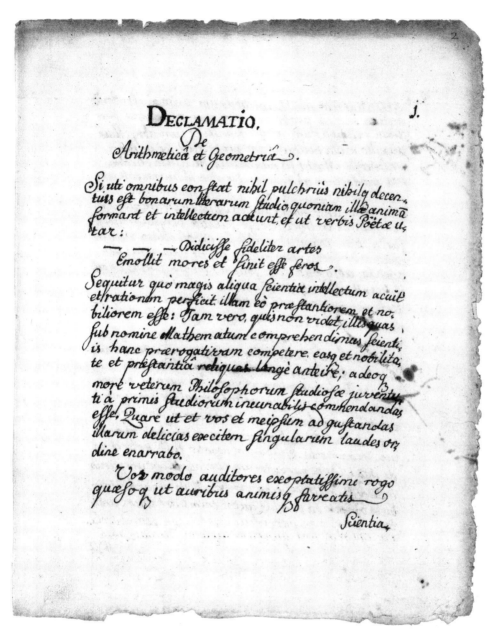

Autograph of the fourteen-year-old LEONHARD EULER: first page of a Latin speech given to his fellow students

RENÉ DESCARTES. Oil painting by
F. HALS, 1648

ISAAC NEWTON. Oil painting by G. KNELLER,
1702

ulty in order to obtain the *prima laurea*, which today would correspond roughly to the "Maturität", the lowest academic degree, which two years later he indeed acquired and put his seal to with a lecture *De Temperantia* [On moderation]. During this biennium he attended the compulsory freshman course of JOHANN BERNOULLI, which included geometry as well as practical and theoretical arithmetic. It was still during this propaedeutical period that the fourteen-year-old EULER gave a speech in Latin to his fellow students with the title *Declamatio: De Arithmetica et Geometria*[19] [Rhetoric: On arithmetic and geometry], in which the youthful author in impressive words not only praised the excellence and usefulness of mathematics for the practical life and the fine arts, but also gave evidence of his being unusually well-read, and of his proficiency in the Latin language. One year later, EULER twice appeared as respondent[20] in public disputations: In January twice about a discourse in logic, then in November about a subject in the history of Roman law.[21] In the fall of 1723 he acquired the academic degree of *magister*, which corresponded to an end of study in the Philosophical Faculty. On this occasion, the new graduate delivered his first public speech (in Latin, of course), in which he compared DESCARTES's

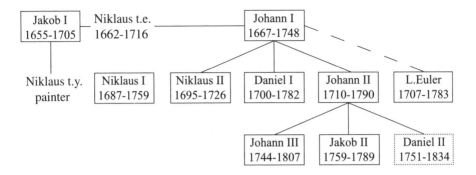

Family tree of the mathematicians BERNOULLI. LEONHARD EULER also finds a place therein as the spriritual son of JOHANN I BERNOULLI. It is true that DANIEL II BERNOULLI has not distinguished himself by publications in mathematics, but nevertheless was sufficiently informed about science that he could occasionally substitute for his uncle DANIEL I in his courses.

and NEWTON's systems of natural philosophy — a theme then extremely timely and till the middle of the century a focal point of interest — because DESCARTES's vortex theory could not be brought in agreement with NEWTON's theory of gravitation and its mathematical implications.[22]

Immediately after his magister exam, LEONHARD EULER enrolled in the Theological Faculty according to the wishes of his father, but his main interest, now as before, was in the higher lectures of JOHANN BERNOULLI to which he gained access not least by virtue of the latter's youngest son, JOHANN II, who became a magister at the same time as he. EULER spent all his free time on mathematical studies, which he conducted with such fervor and success that he aroused the special attention of his teacher and received the privilege of the Saturday *privatissima* described in his autobiography. These — in conjunction with the amicable and scientific contacts with the BERNOULLI sons NIKLAUS II, DANIEL, and JOHANN II — laid the foundation for EULER later to become the most important mathematician of the 18th century.

Concerning JOHANN BERNOULLI's activity as an academic teacher, we have a rather vivid account from one of his university students, although from a somewhat later time. JOHANN JAKOB RITTER (1714–1784) of Bern, who in the summer of 1733 studied in Basel together with JOHANN SAMUEL KÖNIG (1712–1757) and who — though not a novice in mathematics — had to depend on the latter's help in order to be able to follow BERNOULLI's lectures "delivered rather concisely and nervously", tells us [in free English translation from German]:

"Because, to give a brief account of this great man, one has to know that whoever wanted to profitably hear his lectures at that time better had a good grounding in algebra. To him, this subject was far too light to dwell on at length with his usual articulateness. Which is why, in those last years, he gave courses in geometry and algebra only with the greatest resentment. For him, there had to be nothing but 'transcendentals', wherein he lived completely and on which he dwelled at such length until his auditors had a clear understanding of them. He could well tolerate, and liked to see it, when doubts were laid before him, which he resolved with great willingness. And since, moreover, he was in high spirits and could entertain a whole crowd with his clever ideas, he was also very diligent in his lectures and the otherwise burdensome gout would not keep him therefrom . . . . He was very generous (which the fellow citizens of Basel are not usually best known for) and often presented needy auditors with his own lecture fees."[23]

JOHANN BERNOULLI — the undisputed *princeps mathematicorum* after the death of LEIBNIZ (1716) and NEWTON's retreat from the realm of mathematics because of old age — in the course of the Saturday *privatissima*, which had become so famous later on, discovered early on the exceptional talents of the young EULER, and it appears that the Old Master already at that time foresaw in him the even greater master in coming. EULER's first mathematical memoirs (O.II,6; O.I,27) — he wrote them at the age of eighteen resp. nineteen, they appeared in print 1726 resp. 1727 in the *Acta eruditorum* of Leipzig — connect with the ongoing investigations of his great teacher on reciprocal trajectories, and offered him valuable support fire in his long-standing feuds with the English mathematicians. This, BERNOULLI acknowledges in the final lines of his last memoir dedicated to this topic with an almost prophetic-sounding and highly laudatory mention of the young EULER: "Those who wish to pursue this subject further, following the path here indicated, will be able to put their strength to a test by seeking other reciprocal trajectories consisting of lines succeeding one another. That the matter is not impossible can be gathered from what has been achieved by LEONHARD EULER, a young man with the most fortunate talents, from whose cleverness and acuteness we promise ourselves the

Q. F. F. Q. S.

DISSERTATIO PHYSICA

# DE SONO,

QUAM

ANNUENTE NUMINE DIVINO

JUSSU MAGNIFICI ET SAPIENTISSIMI PHI-
LOSOPHORUM ORDINIS

PRO

VACANTE PROFESSIONE PHYSICA

Ad d. 18. Febr. A. MDCCXXVII.

In Auditorio Juridico hora 9.

*Publico Eruditorum Examini subjicit*

LEONHARDUS EULERUS

A. L. M.

*Respondente*

Præstantissimo Adolescente

ERNESTO LUDOVICO BURCARDO
Phil. Cand.

BASILEÆ,

Typis E. & J. R. THURNISIORUM, Fratrum.

Frontispiece of LEONHARD
EULER's "Habilitationsschrift" on
the sound, 1727

greatest, having seen the ease and adroitness with which he penetrated the most secret fields of higher mathematics under our auspices."[24]

This public assessment made by the sixty-year-old grandmaster of the twenty-year-old EULER, in the light of BERNOULLI's character and behavior toward almost all contemporaries — his sons not excluded — is surprising, even sensational. It appears that already at that time BERNOULLI began to consider LEONHARD EULER as his own reincarnation. The salutations in BERNOULLI's letters are characteristic of the respect growing proportionally to EULER's scientific achievements, indeed for the old master's boundless admiration of his student:

1728 (still with fatherly benevolence): "To the highly erudite and ingenious young man"[25];

1729: "To the celebrated and erudite man"[26];

1737 (after EULER solved problems which both elder BERNOULLI, JACOB and JOHANN, could not handle in spite of the greatest efforts): "To the celebrated and by far the most acute mathematician"[27]; and finally

1745: "To the incomparable LEONHARD EULER, the prince among the mathematicians"[28].

With that, the master — at least privately — has surrendered the title to his student. Even if the "Flemish rowdiness" and the pronounced ambition of the often biting, jealous, and envious JOHANN BERNOULLI may have sprouted many an ugly blossom, he nevertheless must be granted the high historical and moral merit of having discovered EULER and having decisively encouraged, patronized, and — above all — tolerated him above himself.

The early period in EULER's activity — his only time in Basel — is further marked by two works which, because of their significance, should not remain unmentioned. He had the audacity to participate with a memoir (O.II, 20) in the public prize competition[29] announced in 1726 by the Paris Academy, namely to determine the optimal way of setting up a mast on a ship — he, "the youthful inhabitant of the Alps"[30], who, other than freighters, ferry boats, and simple canoes on the Rhine river, had never yet caught sight of a ship! Although the first prize was awarded to the then already famous physicist, astronomer, and geodesist PIERRE BOUGUER (1698–1758), EULER's work was cited with an *Accessit*[31], a sort of second prize, which however he had to share with CH. E. L. CAMUS (1699–1768). Highly characteristic of EULER's attitude toward nature is the proud, final paragraph of this work[32]. *I did not find it necessary to confirm this theory of mine by experiment, because it is derived from the surest and most secure principles of mechanics, so that no doubt whatsoever can be raised on whether or not it be true and takes place in practice.*

This almost blind confidence in the rigor of principles and in the *a priori* deductions accompanied EULER to his old age and characterizes a paradigm of his creative work.[33]

With his *Dissertation on the theory of sound* EULER, in the spring of 1727, competed for the physics professorship in Basel which had just become vacant, but — not lastly because of his youth — didn't even make it to the final three, respectively to the lottery[34], even though he could count on the support of the influential JOHANN BERNOULLI. This failure — in retrospect —

was a great fortune, since only in this way could EULER achieve what was denied to his teacher all his life: An arena of work commensurate with his genius and thirst for action — and precisely this he found in the aspiring city of PETER the Great, in the "Venice of the Nord".

# 2
## The first Petersburg period
## 1727–1741

## The establishment of the Academy of Petersburg

At the beginning of the 18th century, events of the greatest importance took place in Russia. Allied with Denmark and Saxony-Poland against Sweden, the Russia of PETER the Great during the Nordic War (1700–1721) against CARL XII fought to gain vital access to the Baltic Sea and secured hegemony in the Baltic area with the victory of Poltawa (1709) and then with the annexation of Livland, Ingria and Finland in the peace treaty of Nystad (1721). An essential preliminary step for this was the citadel built by Tsar PETER I (the Great) with an incredible expenditure of energy at the swampy mouth of the Neva river. From the very beginning, it carried PETER's name as the Russian metropolis and was built according to his own plans — with the aid of many foreign architects, engineers, and technicians — following a strict geometric pattern, starting in 1703 with the colossal bastion "Peter and Paul", and engaging hundreds of thousands of "work slaves".[35]

Of decisive importance for the foundation of an academy in St. Petersburg were the three meetings of PETER with the universal scholar LEIBNIZ in October of 1711 in Torgau, in November of 1712 in Karlsbad, and in June of 1716 in Pyrmont.[36] While the latter's pet plan to cover all of Europe with a network of learned academies had become partially successful with the establishment in 1700 of the "Society of the Sciences" in Berlin, with LEIBNIZ as its first president, further attempts of founding academies in Dresden and Vienna failed — not for lack of interest on the part of the potentates, but at the guillotine of finances. With the tsar, on the other hand, LEIBNIZ's enthusiastic plans fell on fertile ground, and the idea of an academy took on final shape with him — as it seems — after he had attended a meeting of the Academy of Paris on June 19, 1717 during his second European journey.[37] At the time, scientific academies of high and highest niveau existed in London (the "Royal Society"), as well as in Paris,

The building of the Petersburg Academy in the 18th century. Engraving by FICQUET after a drawing by LESPINASSE

GOTTFRIED WILHELM LEIBNIZ. Oil painting by A. SCHEITS, 1703

Rome, Bologna, Florence, and Halle[38], and, when "the Russian Tsar PETER I in 1724 definitively made the decision to (also) found an academy of science in the capital of his empire, St. Petersburg, he thereby crowned his farsighted reform politics. This was aimed at better enabling Russia to subsist within the circle of European powers, without fundamental societal changes, i.e., with the help of absolutism, and under continuing conditions of feudalism. These reforms concerned the state administration, the army, the economy, and the educational system. It needs to be said that aside from the decisive internal factors, which made these reorganizations possible in the first place, there were a number of external factors favorably at work. Russia was able to benefit from experiences gained elsewhere.

This is especially true in the area of science. Faced with the alternative of either developing the required scientific potential for an academy in the country itself through a foreseeably lengthy process, or else call scientists from abroad and entrust them with the task of building an institution of science that met Russia's goals and, at the same time, observed teaching functions, the tsar opted for the latter. Such a procedure was only possible because in Russia itself the possibility for it objectively existed, and because the level of development of the idea of an academy, including the experiences with the organization of scientific academies, made its reception and adaptation a reality. From the beginning, the academy to be founded

Niklaus II Bernoulli. Oil painting by J. R. Huber

Portrait of Daniel Bernoulli as a young man. Oil painting by J. R. Huber

Christian Wolff

Jakob Hermann

was planned as a Russian institution with the scope to developing Russia above all in the areas of science and geography."[39]

On February 2, 1724 Tsar PETER I signed a decree for the foundation of an Academy of Science in Petersburg which, however, he did not live to see but was brought to completion with élan by his widow CATHERINE I. Through the establishment of a scientific center in Russia, the tsar strove for all-around contacts with Western Europe, and he invited the most prestigious scholars of Europe to collaborate at the new academy.[40] Among those were NIKLAUS II and DANIEL I BERNOULLI, who were recommended by CHRISTIAN WOLFF — perhaps in place of their world-famous father JO-HANN I who declined the call — , and they already took their seats at the first assembly of the academy members on November 13, 1725, next to their elder compatriot from Basel, JAKOB HERMANN. WOLFF himself, however, who had been favored by the great tsar as president of the academy, and "who at that time in Halle as head of a new school of philosophy was as popular with the students as he was opposed by his orthodox colleagues . . . , was a prudent man who did not want to get involved in adventures, and when, accused of atheism, he even had to flee from Halle in 1723 within 48 hours, facing 'punishment by hanging', he was most definitely no longer willing to exchange his new refuge in Marburg with the city on the Neva."[41]

## EULER's call – Journey to Russia

As we know from his autobiography, the young LEONHARD EULER would have loved to follow the two other fellows from Basel right away, but of course he had not yet received an official call. Such a call, however, came soon after, on the basis of the recommendations of the two young (and the old) BERNOULLI to the first president of the academy, the imperial physician LAURENTIUS BLUMENTROST (1692–1755). This recommendation was vigorously supported by CHRISTIAN GOLDBACH (1690–1764), the first permanent secretary of the young academy, with whom EULER later, until the latter's death, was to carry on intensive correspondence and also be in close personal contact.[42]

Letter of LEONHARD EULER to JOHANN I BERNOULLI of November 5 (16), 1727. With this letter, EULER inaugurates the correspondence with his teacher, which is to extend over twenty years.

The call to Petersburg came by letter (that has been lost) from BLUMEN-TROST to EULER, and was concretely initiated by a letter in French written by DANIEL BERNOULLI[43] dated September 1726, in which EULER was offered a position as *élève* (later called "adjunct") at a yearly salary of 200 Rbl.. While BERNOULLI admitted in his writing that this actually underestimated EULER's merits, he had succeeded in negotiating with BLUMENTROST considerably better conditions – and indeed, EULER could start his position in June of 1727 with a yearly salary of 300 Rbl. *including free housing, wood, and light*[44], and in addition could still collect 100 Rbl. in Hamburg for travel expenses. He should, so DANIEL BERNOULLI, set out quickly, but should he hesitate because of the cold weather, he should spend the winter in Basel and use the time for studies in anatomy and phsysiology ... This advice was taken by EULER, but for him medicine in Petersburg completely fell by the wayside in favor of the mathematical sciences.

The library on the first and second floor of the Petersburg Academy. Engraving by C. A. WÖRT-MANN after a drawing by G. BON, 18th century

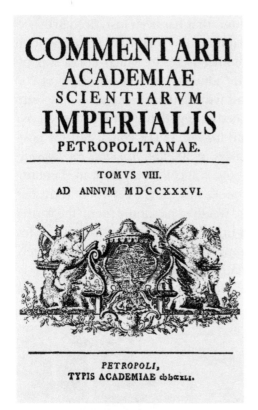

**COMMENTARII**
ACADEMIAE
SCIENTIARVM
**IMPERIALIS**
PETROPOLITANAE.

TOMVS VIII.
AD ANNVM MDCCXXXVI.

PETROPOLI,
TYPIS ACADEMIAE cɔɔcxLɪ.

Frontispiece of the 8th volume of the "Petersburg Commentaries" for 1736, published in 1741. Of the thirteen mathematical memoirs contained in this volume, two are by DANIEL BERNOULLI, the other eleven by EULER.

From his own diary, we have a very detailed account of EULER's journey from Basel to the Russian metropolis.[45] Shortly after his unsuccessful effort for the physics professorship in Basel, and only three days after enrolling in the school of medicine (!), he boarded a ship to Mainz on April 5, 1727, from where he traveled by stage coach via Giessen and Kassel to Marburg, where on April 12 — surely on the recommendation of JOHANN BERNOULLI — he called on the famous philosopher and scientist CHRISTIAN WOLFF[46]. The latter, on April 20, forwarded a letter to EULER, part of which reads:

"I very much regret that you were in such a great hurry, and that I neither was fortunate enough to discuss various matters with you, nor could I give you any courteous assistance to show my high esteem for the Imperial Academy of Sciences, and for the friendship of Mr. BERNOULLI. You are traveling now into the paradise of the scholars (emphasis by

EAF), and I wish hereinafter nothing more than that the Highest above will grant you good health on your journey, and will let you find many years of pleasure in Petersburg. I ask of you to pay my humble respects to his Excellency the President and to give my regards to Mr. BÜLFFINGER, HERMANN, BERNOULLI, MARTINI, LEUTMANN, and also to always remember me kindly ... "[47]

From Marburg the journey continued on land via Hannover and Hamburg to Lübeck, then by ship in stormy weather, which caused EULER to become acquainted with seasickness, past Wismar and Rostock to Reval, which is Tallinn today, from there, on another ship, to the citadel island Kronstadt and finally, after a short ferry trip to the mainland, on foot to Petersburg.

WOLFF was quite correct with his characterization of Petersburg as "paradise of the scholars", for in the early history of the young academy a nearly "golden age" reigned there, thanks to the extremely generous support the young institution was granted under the brief auspices of the Empress CATHERINE I. The construction of two large academy buildings on the shore of the Neva was nearing completion; they contained the art chamber, the library, the observatory, the "anatomic theater", as well as conference rooms and other service rooms. In the beginning, there were public lectures and meetings that used to be conducted in a rather ceremonious manner. Work for installing the academic printing press was still in full swing, while the first academic papers[48] were already being printed. The president of the academy, LAURENTIUS BLUMENTROST, held an influential position at the court, and the letters of the academy members[49] still express the highest expectations. Unfortunately, this was soon to change.

When EULER arrived in Petersburg, there was mourning, consternation and confusion, for one week earlier the Empress CATHERINE I, the widow of PETER the Great, had died at the age of 40 after a reign of only two years, and the power struggles for the succession to the throne were already in full swing.[50] They ended this time with a victory of the old Russians, represented by the lineage of the DOLGORUKI, with the 12-year-old boy PYOTR ALEKSEYEVICH, a grandson from PETER's first marriage, who in 1725 during a palace revolution of the *homines novi* in the style of MENSHIKOV had been passed over in favor of his step-grandmother CATHERINE, now being placed on the tsar throne as PETER II. After the overthrow of the "despised, arrogant Goliath" MENSHIKOV[51] in September of 1727 the reactionary, anti-PETER clan of the DOLGORUKI had the under-age tsar firmly under their control, and thereby also unrestricted sovereign authority. But not for long, for on January 6, 1730, PETER II caught a cold during the festival of the water consecration in Moscow and fell ill with the smallpox, which led to his death on the day — January 19, 1730 — that should have been his wedding day to the princess ANASTASIA DOLGORUKAYA. With him, the male lineage of the ROMANOVS became extinct.

What turned out to be an unfortunate circumstance for the academy was the fact that with PETER's II succession to the throne, the court — and along with it also his personal physician, the academy's president BLUMENTROST — were transferred to Moscow. In this way, the administration of the academy — although not *de jure*, but *de facto* — fell into the hands of the librarian and chancellor JOHANN DANIEL SCHUMACHER (1690–1761) who, to be sure, displayed drive and administrative skills, but with whom no one could get along for long. Almost all members of the academy — most of all the older ones — opposed the arbitrariness of this politically savvy opportunist, who, directly or indirectly, caused the departure of several of the most important academicians such as DANIEL BERNOULLI, HERMANN, and BÜLFINGER. The one who probably had to suffer most under the petty despot SCHUMACHER was the great MIKHAIL LOMONOSOV, as is so painfully documented in his correspondence with EULER[52]. The decade of ANNA IVANOVNA's reign (1730–1740) brought some relief to the academicians insofar as the new empress undertook a restoration along the lines of PETER the Great (or rather had it brought about), and moved the residence back to Petersburg, yet SCHUMACHER — despite several changes in the presidency — remained the strong man in the academy.[53]

# The first scholars at the Academy of Petersburg

Let us take a quick look at the colorful, motley crowd of (non-Russian) scholars who populated the Petersburg Academy in its first years.[54] After CHRISTIAN WOLFF as well as JOHANN BERNOULLI had declined to accept the call and had themselves confirmed only as foreign (honorary) members of the academy, the latter — upon the suggestion of WOLFF — sent 1725 his two (unmarried) sons NIKLAUS II and DANIEL to Petersburg, and the BERNOULLI clan managed to get both brothers appointed as professors. DANIEL BERNOULLI, who was to move up to the *élite* of physicists in his century, worked for eight years in Petersburg, but his brother NIKLAUS to whom he was very close, his father's favored son, died there from an intestinal ulcer after only nine months[55]. A man of the first hour was also the important mathematician from Basel, JAKOB HERMANN[56] (1678–1733) — likewise mediated by WOLFF — who as nestor of the "Basel delegation" drew the highest salary of 2,000 Rbl. annually, but with his almost fifty years must already be counted among the seniors, considering the average age of the first Petersburg academicians. EULER's entrance completed the "Basel quartet", which was to write a significant piece of history of the Petersburg Academy. Purely numerically, however, the land Württemberg exceeded the Basel contingent by quite a bit. One of the most brilliant representative of these "seven Swabians" who went to the Petersburg Academy doubtlessly was the widely educated GEORG BERNHARD BÜLFINGER (1693–1750). A leading theologian of his time, and philosopher in the vein of WOLFF, with whom among others he also studied, he occupied himself intensely with mathematics, physics, and botany, and was considered an expert on the science of fortification. In 1725, he took on the professorship for logic, metaphysics, and physics at the Petersburg Academy, and returned in 1731 to Württemberg, where under the Duke CARL he assumed high political positions and managed to move up as far as prime minister. In BÜLFINGER's tow were the philosopher CHRISTIAN FRIEDRICH GROSS (1698?–1742), who in connection with the throne revolt of 1741 took his life in Petersburg on New Year's day of the following year[57], and the mathematician and astronomer FRIEDRICH CHRISTOPH MAYER (1697–1729), who a few years later was to succumb to consumption. Also coming from Tübingen, the excellent anatomist JOHANN GEORG DUVERNOY (DUVERNOIS) (1691–1759) from Montbéliard (Mömpelgart, then belonging to Württemberg) joined the Academy. At one time he studied medicine in Basel among other places, completed his studies in Paris, and later in Tübingen was the academic advisor of ALBRECHT VON HALLER. DUVERNOY, too, brought with him two able young men: the mathematician JOSIAS WEITBRECHT (1702–1744), who soon, however, changed to anatomy and became an acknowledged master in this subject, and the mathematician and physicist GEORG WOLFGANG KRAFFT (1701–1754), who may be considered one of EULER's closer friends in the first period of Petersburg[58], and who next to him was the most active — and often even the only — collaborator in the mathematical-physical class of the *Commentarii*. JOHANN GEORG GMELIN (1709–1755), the "Benjamin" of the group, also came from DUVERNOY's and BÜLFINGER's "school of Tübingen", and arrived in Petersburg two months after EULER. He was the precocious, highly talented son of his equally named father, a famous pharmacist and chemist. The young botanist and geologist GMELIN subsequently was to gain fame with the four-volume monumental work on the flora of Siberia, which he presented to the public as a kind of research report on the (so-called second) Kamchatka-expedition of Captain BERING, which lasted almost ten years and in which he participated.[59]

From other German districts there were still five more personalities that joined the Petersburg Academy. The historian GOTTLIEB SIEGFRIED BAYER (1694–1738) came as Professor for Greek-Roman archaeology, and the theologian JOHANN GEORG LEUTMANN (1667–1736), the senior of the group, was called upon the suggestion of HERMANN as mechanic and optician after having given up his country parish in Wittenberg. Of greater significance for EULER's activity, however, was the presence of three other celebrities, namely JOSEPH-NICOLAS DELISLE (DE L'ISLE) (1688–1768), GERHARD FRIEDRICH MÜLLER (1705–1783), and CHRISTIAN GOLDBACH (1690–1764).

The astronomer and geograph DELISLE, invited personally by PETER the Great, reported for duty from Paris with a large assemblage of instruments to establish the Petersburg Observatory. He was a very able experimental observer and excellent theoretician, who later was entrusted with the directorate of the "Geography Department" of the Academy which he founded himself. In these functions he was in a certain sense EULER's superior and teacher during the first years. DELISLE is considered to be the actual founder of the "Petersburg astronomical school", and EULER owes much to this important scholar with regard to the rigorous formalization of spherical trigonometry, the modern conception of celestial mechanics, and the foundation of mathematical cartography[60]. DELISLE left Russia in 1747 and committed a serious breach of trust, as he took with him important research material of the Kamchatka-expedition and made it public in

Paris — very fragmentarily and imprecisely — still before it was published in Russia itself. Later on, this was to lead to spectacular controversies, which also EULER — then already in Berlin — was drawn into[61].

To EULER's intimate circle of friends also belonged the historian and geograph GERHARD FRIEDRICH MÜLLER, who came to Petersburg two years before EULER and until 1728 had to teach Latin, history, and geography at the "Academic Gymnasium"; subsequently he was recruited by SCHU-MACHER, the managing director of the Academy, for organizational work and for editing two newspapers in Petersburg. After having fallen out with SCHUMACHER (1731), MÜLLER turned entirely towards history and geogra-phy, which found expression in the first parts of the *Collection of Russian history*, published in 1732/33. This collection contains a vast amount of documents and reports on the history of Russia — until 1764 altogether nine volumes were edited by MÜLLER. In 1733, MÜLLER — like GMELIN — joined the second Kamchatka-expedition and subsequently wrote his ex-tensive four-volume opus *History of Siberia* (*Istoriya Sibirii*), of which, how-ever, he lived to see only the publication of the first volume (1750) and partial publication of the second. Later, MÜLLER was appointed historian of the Russian empire and rector of the Academic University, and in 1766 at-tained the position of trust as director of the archive of the Council for For-eign Affairs in Moscow, without, however, having to relinquish the close tie with the Academy in Petersburg. The correspondence of EULER with MÜLLER embraces the period from 1735 to 1767 and is a primary source not only for the relations between the Berlin and Petersburg Academy and for their internal conditions, but also for the biography of the two corre-spondence partners[62]; in our time it was completely edited in the original language (German)[63].

EULER, it appears, most cordially attached himself — apart from DANIEL BERNOULLI, whose dwelling he shared during the first six years — to the diplomat and mathematician CHRISTIAN GOLDBACH (1690–1764)[64]. The latter, without doubt, was one of the most remarkable and universally ed-ucated personalities of the young congregation[65]. Born the son of a min-ister in Königsberg, he there took up the study of law, which in 1712 he completed with the licentiate of the Dutch university in Groningen. His extensive travels through many countries of Europe helped him to get his proverbial worldliness and extraordinary proficiency in languages. In addition, they brought him personal acquaintance and correspondence with leading personalities in science like LEIBNIZ, WOLFF, and the three brothers BERNOULLI. In spite of his connections with BLUMENTROST he was not called as academician to Petersburg, yet he appeared there uninvited, and in September of 1725 he received a contract for five years as a member of the Academy; he became the first permanent secretary of this institution, to the organization of which he contributed significantly. It is today also established that he played a decisive role in the call of EULER. Even though GOLDBACH is known to most of today's mathematicians only by virtue of a theorem (more precisely: a conjecture) named after him[66], he should, as a mathematician, not be dismissed with disrespectful handwaving (since "merely an amateur"). His correspondence with EULER[67] alone — it extends over 35 years and comprises close to 200 letters — already is a jewel in the history of science of the 18th century and proves GOLDBACH to be a mathematical instigator of the first order.

The sessions of the academy, which normally took place on Tuesdays and Fridays starting at 4 p.m., it appears, were conducted in a rather lively, even fiery, manner. This had to do not only with the colorfully thrown together group of members, but also with the sharp contrast between the LEIBNIZian rationalism, as represented by the "WOLFFians" under the leadership of BÜLFINGER, and the English empiricism of NEWTON's observance, which DANIEL BERNOULLI and EULER, among others, spoke up for, whereas JAKOB HERMANN held together with the older generation of the BERNOULLIS and took side with the WOLFFians. In view of the seven Swabians and the four from Basel, one can easily imagine how the sometimes sharp disputes were carried on in an amusing jumble of Latin and Alemannic; some small samples of such discussions, with sometimes embarrassing consequences, are described by SPIESS[68], and for the discussions about tolerance and enlightenment in the Russia after PETER in general, one may consult the extensive monograph of WINTER[69].

# EULER's first years in the tsardom

As shown by his diaries, EULER — contrary to most foreign colleagues — began immediately to study the Russian language in order to be able to express himself freely, both in writing and speaking. He adjusted quickly and easily to the new conditions of life and took an active part in the manifold activities of the Academy. The latter — as an important state institution — was charged with, among other things, the training of a national scientific cadre at the "Gymnasium" and the university of the Academy on the one hand, and on the other hand, with the solution of diverse technical problems and with carrying out commissions of the government to study the Russian empire and its natural resources.

Official documents attest to the fact that EULER for several years gave courses in mathematics, physics, and logic, and participated significantly in examinations for the cadet corps. He examined the masters and geodesists entering the Academy, and participated in various expertises, as for example on the examination of the weights in the Academy's Office of Weights and Measures, on the council of commerce and the customs department of Petersburg, on the sawmill and the fire brigade created by A. K. NARTOV; following EULER's design, a steam engine was also built using PAPIN's principle. Next to such strenuous duties, there remained still sufficient time to the young scholar for his principal and favored occupation, namely for the mathematical sciences. Already the second volume of the *Commentarii* (1727) of 1729 shows three papers of EULER: a purely mathematical one on reciprocal trajectories[70], the second on tautochrones[71], and a third, physical one, on the elasticity of air[72]. It was with a scientific communication (on a question of hydraulics) that EULER on August 5, 1727, addressed the Petersburg Academy for the first time. The hydro-mechanical and -dynamical studies, though, which occupied him there already in the first years, he put on the back burner in deference to his friend DANIEL BERNOULLI, who still in Petersburg drafted a first version of his *Hydrodynamics*[73], which was to become a classic.

When JAKOB HERMANN in 1730 returned to Basel, DANIEL BERNOULLI took over the latter's professorship for mathematics, and EULER in 1731 was given the physics professorship, which had become vacant after BÜLFINGER's departure. At the same time he advanced to an ordinary member

Last page of a letter, written in Russian, by LEONHARD EULER to NARTOV, June 18, 1743 (old style)

of the Petersburg Academy. Two years later, when DANIEL BERNOULLI returned to his native country, EULER took over — finally! — the now vacant professorship of mathematics and left the physics chair to his friend and colleague KRAFFT.

His yearly salary of 600 Rbl. now allowed EULER to think of marriage, all the more so as the commander of the cadet corps, Baron VON MÜNNICH, proposed to him *to hold classes in the cadet corps and at the same time give examinations to the instructors*[74], which came with an increase in salary of 400 Rbl. LEONHARD's choice fell on KATHARINA GSELL, a daughter of the second (divorced) marriage of the artist GEORG GSELL (1663–1740) originating from St. Gallen, whom the tsar PETER I had come to know at one time in Holland and whom he had engaged for his art academy. Also KATHARINA's stepmother, GSELL's third wife, worked as painter at the same institute; she was a daughter of the painter and natural scientist MARIA SIBYLLA MERIAN (1647–1717), still famous today, who made a name for herself through her studies of insects in Surinam and whose portrait adorned, as is known, the 500 DM-banknote (her father was the important engraver and topographer MATTHÄUS MERIAN [1593–1650] from Basel).

The wedding took place on December 27, 1733 (January 7, 1734); the EULERS moved into a house of their own in the Tenth Line of the Wassiljevski-Island, which, built entirely of wood, was *über die Maßen wohl conditionirt*[75] [exceedingly well furnished[75]] and was located near the academy building on the bank of the Neva. EULER's wife, "who was to make this house a home for him, was born in Amsterdam in the same year as he himself — which is about all we know about her. Since neither EULER nor anybody else speaks about her, she must have been (according to a known proverb) a fine woman! — Apparently, the marriage was a happy one."[76]

On November 16 (27), 1734, she gave birth to her first son, who received the names JOHANN ALBRECHT after the first godfather KORFF. The second godfather was, as already mentioned, CHRISTIAN GOLDBACH. "The fact that the Chamberlain KORFF, highly respected at the court, who only shortly before, on September 18, 1734, had become president of the Academy, was godfather of his [EULER's] son, together with the then permanent and influential member of the Academy of Sciences, GOLDBACH, bears witness of the high esteem in which EULER was held at the Petersburg Academy already at this time."[77]

How the young mathematician might have looked like around this time is shown in a print of V. SOKOLOV (Fig. opposite the title page) after a (lost) painting of J. G. BRUCKER. Of EULER's wife KATHARINA, unfortunately, no portrait is extant[78], but from a letter of DANIEL BERNOULLI we know that one — together with a portrait of EULER, which today is also lost — has been shipped to Basel: *"Die gemählde haben wir endlich empfangen ... Ew. HEdgb. und dero Fr. Liebsten portraits sehen sehr gleich."*[79] [The paintings we finally received ... Your Honorable and your dear wife look very true to life.[79]]

In the first months of the year 1735, EULER was struck by a severe blow: a life-threatening illness, whose nature today can no longer be exactly diagnosed. But it appears that this illness, which according to contemporary accounts was to have manifested itself with a "fiery fever", has had an inner connection with the one in the late summer of 1738, which cost EULER his right eye[80]. The seriousness of that first illness becomes clear from a letter to EULER which his friend DANIEL BERNOULLI begins with the following words: *Allervorderst gratuliere ich Ew. HEdgb. zu dero wieder so glückl. erlangten gesundheit und wünsche von hertzen eine lange continuation derselben. Wie mir Hr. MOULA schreibt, so war nicht nur jederman bej ihrer kranckheit umb Sie bekümmert, sondern auch sogar ohne hoffnung Sie wiederumb von derselben restituirt zu sehen. Es ist gut, daß weder ich noch dero Eltern eher etwas darumb gewust, als man dero völlige genesung vernommen. Es hat sich sonderlich auch der 'orbis mathematicus' über dero wunderbahre genesung zu erfrewen.*[81] [First of all, I congratulate you on your so fortunately restored health and wish you from my heart a long continuation thereof. As Mr. MOULA writes to me, not only did everyone worry about you because of your illness, but even had no hope to ever see you recovered therefrom. I am glad that neither I nor your parents knew anything about it before receiving the news of your complete recovery. In particular, also the mathematicians of the world have to be very pleased about your wonderful recovery.[81]]

Alas, the "continuation" of Euler's health did not last very long: Subsequently, the same (?) violent infectious disease led to the loss of the right eye (1738), as is clearly evident in all later portraits of the mathematician. Thus, the portrait of BRUCKER/SOKOLOV is the only one of EULER which shows him with two (relatively) healthy eyes.

EULER's own concealment of his illness of 1738 — even from his parents — may (after R. BERNOULLI) be due to a reverent feeling of affection, but from another letter of DANIEL BERNOULLI to EULER it can be surmised that the latter, probably in October of 1738, had informed his parents of a serious eye trouble endangering the health of the eyeball[82]: *Dero Hr. Vatter wird ihnen vielleicht gemeldet haben, wie starck mir Dero betrübter zufahl zu hertzen gegangen: Gott wolle Sie von fernerem unglück behüten; wir hätten gar gern eine genawere beschreibung ihrer kranckheit gehabt: ob der bulbus oculi gantz verderbt und die humores ausgerunnen, oder ob dem äußeren ansehen nach, der bulbus noch unversehrt seje.*[83] [Your father may have informed you, how much your sad affliction has worried me: may God protect you from further misfortune; we would have much liked to have had a more precise description of your illness: whether the eye ball is completely damaged and the fluid leaked out, or whether from the external appearance the eye ball is still intact.[83]]

Unfortunately, EULER's reply to this letter is missing, but here is the place to emphatically banish into the realm of legends the fairy tales and errors that have been circulating forever about the cause of the loss of the eye:

1. In his impressive and now famous eulogy ("Lobrede") on LEONHARD EULER, the then twenty-eight-year-old NIKLAUS FUSS from Basel, who during the last ten years of EULER's life, together with JOHANN ALBRECHT EULER, was his closest assistant, reports (1783) the following: "About his iron diligence, he [EULER] gave a still more remarkable example when in 1735 a calculation[84] needed to be done, which was urgent and for which various academicians wanted to have several months' time, and which he completed within three days. But how dearly did he have to pay for this effort! It inflicted on him a fiery fever, which brought him at death's door. His nature, though, prevailed and he recovered, but under loss of the right eye, which was taken away by an abscess which developed during the illness."[85]

In this description, not only the connection between cause and effect is wrong, even impossible, but also the point in time is off by three years.

2. On August 21, 1740, EULER wrote to CHRISTIAN GOLDBACH, who also stayed in St. Petersburg: *Die Geographie ist mir fatal. Ew. Hochedelgeb. wis-*

*sen, daß ich dabei ein Aug eingebüßt habe; und jetzo wäre ich bald in gleicher Gefahr gewesen. Als mir heut morgen eine Partie Karten, um zu examinieren, zugesandt wurden, habe ich sogleich neue Anstöße empfunden. Dann diese Arbeit, da man genötiget ist, immer einen großen Raum auf einmal zu übersehen, greifet das Gesicht weit heftiger an, als nur das simple Lesen oder Schreiben allein. Um dieser Ursachen willen ersuche ich Ew. Hochedelgeb. gehorsamst, für mich die Güte zu haben, und durch Dero kräftige Vorstellung den Herrn Präsidenten dahin zu disponieren, daß ich von dieser Arbeit, welche mich nicht nur von meinen ordentlichen Funktionen abhält, sondern auch leicht ganz und gar untüchtig machen kann, in Gnaden befreiet werde ...* [86] [The geography is fatal to me. Your Honorable knows that because of it I lost an eye; and now I would soon have faced the same danger. When this morning a series of maps were sent to me for examination, I felt immediately new attacks. Because this work, in which one is forced to always survey a large area at once, strains the vision much more vigorously than simple reading or writing alone. For these reasons, I request Your Honorable most obediently to do me the favor to dispose the president through strong intervention to graciously release me from this work, which not only prevents me from attending to my ordinary functions, but also easily can render me utterly unfit ... [86]]

As a matter of fact, since the beginning of the thirties, EULER had to do excessively much for *geography* (Kamchatka-expedition, cartography of Russia), and since 1740 he was even responsible for the entire Department of Geography, but EULER here falls victim — like many patients still today in analogous situations — to an erroneous notion, as R. BERNOULLI proves convincingly, because a stress situation can at most have a triggering effect with regard to general illnesses.

# First principal works

In spite of the serious setbacks in his health, suffered in the years 1735 and 1738, EULER, with incomprehensible diligence, moved the fronts of several scientific disciplines simultaneously: During the "first Petersburg period" over fifty memoirs and books written by him appeared in print.[87] These works can be divided into fourteen research areas, which are listed here only by keywords: algebra (theory of equations); number theory (prime numbers, Diophantine analysis); arithmetic; geometry (topology); differential geometry (reciprocal trajectories, geodesics); differential equations; theory of series (infinite series); calculus of variations; mechanics (global representation, tautochrones, theory of impact and elasticity); theory of ships; physics (elasticity of air, the nature of fire); astronomy (positional astronomy, planetary orbits); theory of tides; and music theory.

In the following subsections, EULER's principal works on mechanics and on naval theory are briefly characterized, but more space will be devoted to music theory in view of the certainly wider general interest.[88]

## Mechanics*

In the perhaps best global, concise German presentation of EULER's achievements in the area of mechanics in the widest sense, MIKHAILOV[89] stresses that for EULER, mechanics was the first serious passion, as can be clearly seen in his (extant) notebooks and diaries which he kept at the age of eighteen and nineteen. In 1736 there appeared, in two volumes, as a supplement to the *Commentarii*, EULER's first great masterpiece, the *Mechanica*, which represents a milestone in this branch of science, and which also immediately earned him high recognition among the scholars of the time. In the Introduction to the first volume, EULER sketches a comprehensive program for this discipline, whose main feature is the systematic and fruitful application of analysis, that is, of the differential and integral calculus, to the then existing as well as new problems of mechanics. The predecessors of EULER proceeded — summarily spoken — in a synthetic-geometric fashion, for which an outstanding example is the immortal *Principia mathematica* of NEWTON[90]. Also JAKOB HERMANN in Basel, in spite of the modernity striven for in the *Phoronomia*[91], was unable to break away from the baroque style of JAKÓB BERNOULLI, his former teacher. EULER already here — as also later in the optics — proceeds consistently in an analytic manner and demands for mechanics uniform analytic methods which should lead to clear and direct statements and solutions of the relevant problems. The title of

# MECHANICA
SIVE
## MOTVS
# SCIENTIA
ANALYTICE
EXPOSITA
AVCTORE
## LEONHARDO EVLERO
ACADEMIAE IMPER. SCIENTIARVM MEMBRO ET
MATHESEOS SVBLIMIORIS PROFESSORE.

## TOMVS I.

INSTAR SVPPLEMENTI AD COMMENTAR.
ACAD. SCIENT. IMPER.

◄▓▒▒▒▓▓▓▓▓▓▓▓▓▓▓▓▓▓▓▓▓▓▓▓▓▓▓▓▓▓▓▓▓▓▓ ►

PETROPOLI
EX TYPOGRAPHIA ACADEMIAE SCIENTIARVM.
A. 1736.

Frontispiece of EULER's "Mechanica",
Petersburg 1736

the work already contains the whole program: *Mechanics or the science of motion described analytically*[92].

EULER begins with the kinematics and dynamics[93] of a mass point and in the first volume treats the free motion of a mass point in vacuum and in a resisting medium. The section on the motion of a mass point under the action of a force directed toward a fixed point is a brilliant analytic reformulation of the corresponding chapter in NEWTON's *Principia*. In the second volume, EULER studies the forced motion of a mass point, and in the context of the equations of motion of a point on a given surface, solves a series of differential-geometric problems in the theory of surfaces and geodesics. Almost thirty years later, in the *Theoria motus*[94], the so-called "Second Mechanics", EULER gave a new exposition of point mechanics, in which he projects the force vectors — after the model of COLIN MACLAURIN[95] — onto a fixed, orthogonal system of coordinates in three dimensions and, in connection with the investigations of rotational motion, derives the differential equations of dynamics relative to a system of principal axes, which characterize these motions. He furthermore formulated the law of motion, expressible in terms of elliptic integrals, of a rigid body around a fixed point ("EULERian angles"), to which he was led in his study of the precession of the equinox and the nutation of the axis of the earth. Other cases of the theory of tops, in which the differential equations are integrable, were later discovered and treated by JOSEPH LOUIS LAGRANGE (1788) and SOFIA KOVALEVSKAYA (1888), a pupil of WEIERSTRASS.[96]

## The theory of ships*

In the domain of hydromechanics[97], EULER's first great work was his two-volume opus on "Naval science", the *Scientia navalis*[98]. This work, which was completed already ten years before its publication, as a supplement to the *Novi Commentarii*, "represents after the *Mechanica* ... the second milestone in the development of rational mechanics and to this day has nothing lost in importance. Here, indeed, not only are for the first time postulated in perfect clarity the principles of hydrostatics, and based on this, a scientific foundation given for the theory of shipbuilding, but the circle of topics taken up here provides us with a synopsis of almost all relevant lines of development of mechanics in the 18th century."[99]

In the first volume, EULER treats the general theory of equilibrium of floating bodies and studies — then a novum — problems of stability as well as small vibrations (rockings) in the neighborhood of the state of equilibrium. In this connection, EULER defines via the (directionally independent) fluid pressure an *ideal fluid*, which later served CAUCHY as a model for the definition of the stress tensor. The second volume brings applications of the general theory to the special case of ships.[100] With the *Scientia navalis*, EULER founded a new branch of science, as it were, and exercised a lasting influence on the development of seafaring and ship engineering. It is known only to a few specialists that we owe EULER to a large degree the technically feasible principle of impeller drive and of the naval screw. Naturally, these bold projects in EULER's time were dismissed as working only in theory, since the propulsion energies needed for their realization were not yet available. Well known in the history of technology, however, are EULER's experiments on SEGNER's water-powered machine[101] and his related theory of the water turbine. JAKOB ACKERET (1898–1981) in 1944 had such a turbine built as a prototype, following EULER's precepts, and observed that

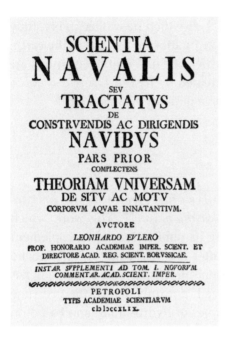

Frontispiece of EULER's "Scientia navalis"
("Theory of ships"), Petersburg 1749

Prototype of a water turbine built according to EULER's proposals

the efficiency of EULER's machine was more than $71\%^{102}$ — a sensational result considering that today, with the most modern means available and for comparable dimensions, the efficiency achievable for such a turbine is just a little over 80%.

In the early fifties falls the composition of a few truly classical memoirs on an analytic theory of fluid mechanics, in which EULER developed a system of fundamental formulae for hydrostatics as well as hydrodynamics. Among them are the equations of continuity for fluids of constant density, the equation — usually named after LAPLACE — for the velocity potential, and the general "EULER equations" for the motion of ideal (thus frictionless) compressible and incompressible fluids. Characteristic also for this group of papers is the derivation and application of certain partial differential equations governing problems in this area. As we know from autotestimonials, EULER thought especially highly of these things — and rightly so.

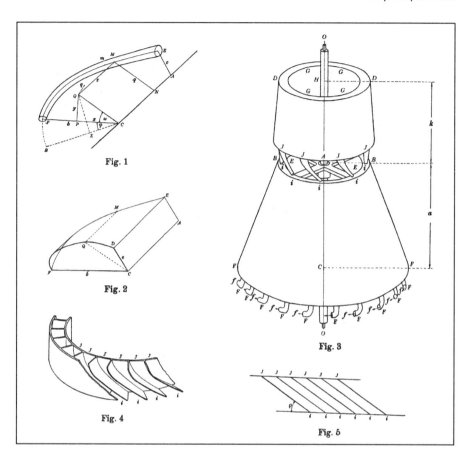

Figures from EULER's memoir of 1754 on water turbines

## Music theory*

Music has also occupied EULER since his youth, as is shown by his note-books from the time in Basel, and already in about mid-1731 he was able to deliver to the Academy of Petersburg — in a certain sense as a "mandatory assignment" — a music-theoretical work, which according to the first proofs carried the title *Tractatus de musica*, but only eight years later appeared in print in Petersburg as a monograph under the title *Tentamen novae theoriae musicae* ... [Attempt of a new theory of music].[104] This work, which below we call briefly *Tentamen*, was later followed up by Euler with yet three other memoirs[105], mainly dedicated to the natural seventh.

In the *Tentamen*, Euler treats not only the mathematical laws of consonance, but also aspects of the theory of composition. On the basis of old-Pythagorean principles of harmony, according to which the subjectively perceived degree of beauty of an interval depends on the simplicity of the frequency ratio, Euler — in the first attempt undoubtedly influenced by his predecessors Mersenne, Descartes, and Leibniz — operates with his number-theoretically motivated concept of *degree of consonance*[106]. The fact that Euler's deductive theory, contrary to Rameau, did not get accepted in its logical rigor can certainly not be blamed only on the equal temperament[107], spreading and accommodating certain practitioners already at that time, but rather on the fact that Euler's tonal system was not even "well tempered" in the sense, say, of Werckmeister, and did not sufficiently meet the needs of contemporary musical practice. Moreover, Euler's *Tentamen* to musicians may well have appeared too mathematical, and to mathematicians too musical.

As a second pillar, Euler places his *substitution theory* next to the *gradus theory*. It is a sort of "theory of properly hearing" as an answer to the question what modifications occur in a perceiving subject when it perceives differently than it actually should perceive according to the objective, i.e. physical, conditions. According to this conception the natural bounds existing in the affection of the senses is allowing the active state of consciousness room to move, within which the acoustic sensations received simultaneously or consecutively are absorbed into structures which more or less differ from what is written in the score or produced on an instrument. H. R. Busch[108] has shown that this idealization process provides an explanation for known phenomena such as conceptual dissonance, pivotal function, and enharmonic change. In this sense, Euler later uses the natural seventh (4 : 7) to explain the surprising consonance of the dominant seventh chord. According to Vogel[109], modern experimental investigations of nine researchers are supposed to have clearly demonstrated the high consonance and the blending capability of the seven group.

Euler's tonal system runs more or less in parallel with the temperament of Kirnberger, having the goal to keep as pure as possible the natural harmonic basis resp. the *genus diatonicum*. Let us present this in Euler's

First page of EULER's letter to JOHANN I BERNOULLI of June 20, 1740. The crossings out of the lines 5–9 from above were done (probably) by JOHANN II BERNOULLI. Similar deletions can be found in most of EULER's letters to JOHANN I BERNOULLI, almost always when there is talk about financial matters.

own words from a letter to JOHANN I BERNOULLI[110]: *At the beginning of next year my memoir on music*[111]*, which I had written a few years ago, will now also go into print, in which, so I think, are revealed the true and inherent principles of harmony. This theory in fact shows quite specifically the agreement of the old music with the contemporary one. Namely, it is to be shown how a system of all different, related tones which produce a certain harmony can be brought under a certain general term, whose individual divisors generate precisely the tones of the system. Thus, for example, the general term $2^n \cdot 3^3 \cdot 5$ is the 'exponent' of the PTOLEMAIC tonal system, since all its divisors within the ratio 1 : 2 yield the tones of this system and fill the interval of a single octave. The simple divisors — disregarding the powers of two, which, after all, only raise the tone by one or several octaves — are the following:*

$$1; \quad 3; \quad 3^2; \quad 3^3; \quad 5; \quad 3\cdot 5; \quad 3^2 \cdot 5; \quad 3^3 \cdot 5.$$

*The individual terms, multiplied by powers of two in such a way that they fall into the next octave, yield for the consecutive tones of the diatonic system*[112] *the numbers*

$$96 : 108 : 120 : 128 : 135 : 144 : 160 : 180 : 192$$
$$C \ : \ D \ : \ E \ : \ F \ : \ F^\sharp \ : \ G \ : \ A \ : \ B \ : \ c.$$

*This system differs from the usual one only by the here inserted tone $F^\sharp$, whose sole deletion would in no way upset the theory. The exponent of the nowadays most frequently used diatonic-chromatic system of 12 tones within an octave is, as I have observed, $2^n \cdot 3^3 \cdot 5^2$, and its 12 simple divisors are*

$$1; \quad 3; \quad 3^2; \quad 3^3; \quad 5; \quad 3\cdot 5; \quad 3^2 \cdot 5; \quad 3^3 \cdot 5; \quad 5^2; \quad 3\cdot 5^2; \quad 3^2 \cdot 5^2; \quad 3^3 \cdot 5^2.$$

*If one reduces them by means of powers of two to the interval of a single octave, they yield the following tonal system:*

$$2^7 \cdot 3 : 2^4 \cdot 5^2 : 2^4 \cdot 3^3 : 2 \cdot 3^2 \cdot 5^2 : 2^5 \cdot 3 \cdot 5 : 2^9 \cdot 1 : 2^2 \cdot 3^3 \cdot 5 : 2^6 \cdot 3^2 :$$
$$384 \ : \ 400 \ : \ 432 \ : \ 450 \ \ : \ 480 \ \ : \ 512 \ : \ 540 \ \ : \ 576 \ :$$
$$C \ : \ C^\sharp \ : \ D \ : \ D^\sharp \ \ : \ E \ \ : \ F \ : \ F^\sharp \ \ : \ G \ :$$

$$2^3 \cdot 3 \cdot 5^2 : 2^7 \cdot 5 : 3^3 \cdot 5^2 : 2^4 \cdot 3^2 \cdot 5$$
$$600 \ \ : \ 640 \ : \ 675 \ : \ 720$$
$$G^\sharp \ \ : \ A \ : \ B^\flat \ : \ B.$$

*And these tone ratios agree just as exactly with those which have been firmly introduced by the musicians, only the tone B♭ — the only one — deviates just a little bit. They in fact normally set the ratio to A : B♭ = 25 : 27, whereas the theory gives for it 128 : 135. Since, however, the whole tonal system can be expressed by an 'exponent', an arbitrary consonance can in this way be represented by the exponent and the degree of pleasure of the consonance determined through it. All this I have fully described and proved in the short treatise, which will soon come out.*

JOHANN BERNOULLI did not let himself into any discussion, and in March of 1739 he wrote only a few noncommittal words on this matter: "In der Musik bin ich nicht sehr bewandert, und die Grundlagen dieser Wissenschaft sind mir zuwenig vertraut, als daß ich Ihre diesbezüglichen Entdeckungen beurteilen könnte. Das, was Sie in Ihrem Brief — wenngleich nur flüchtig — berühren, scheint wirklich hervorragend zu sein. Doch wenn ich Ihre Abhandlung selbst gesehen haben werde, die Sie über die Prinzipien der Harmonie veröffentlichen wollen, hoffe ich, daß mir daraus ein helleres Licht aufleuchtet zu tieferer Einsicht in die Vortrefflichkeit Ihrer Entdeckungen."[114] [In music I am not very versed, and with the foundations of this science I am not sufficiently familiar that I could judge your discoveries in this area. What you touch on in your letter — if only in a cursory manner — seems to be really outstanding. But once I myself will have seen your memoir, which you want to publish on the principles of harmony, I hope that a brighter light will shine from it for a deeper insight into the excellence of your discoveries.[114]]

It must be said that JOHANN BERNOULLI, also in later years, has not returned to this anymore. DANIEL, on the other hand, who of course has also read EULER's letter to his father, wrote to his friend (with the father's mail to Petersburg) rather frankly, with a shot of scepticism: *Dero opus musicum wird auch sehr curios sejn: doch aber zweiffle ich daran, ob die musici dero temperatur wurden annemmen; daß der terminus generalis $2^n \times 3^m \times 5^p$ alle tonos fere, ut sunt recepti, gebe, ist vielleicht nicht anderst als eine observation zu betrachten. In der music glaub ich nicht, daß am meisten auff eine harmonia perfectissima reflectiert werde, weilen man doch mit dem gehör ein comma nicht distinguieren kan. Gesetzt die progressio geometrica[115] gebe die tonos so accurat, daß dieselbe eine proportionem simplicem nicht zwar accurat sondern nur quoad sensum accurat geben, so wurde dieselbe zu praeferieren [sejn], wegen*

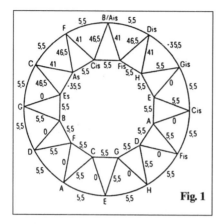

**Fig. 1**

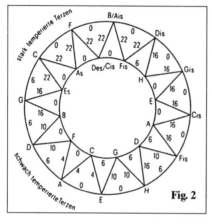

**Fig. 2**

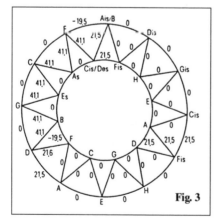

**Fig. 3**

**Fig. 1** shows the mediant temperament of the Renaissance, which for a long time—on organs sometimes well into the 19th century—was, with variations, the most common temperament for keyboard instruments. It is obtained through the sequence of eleven fifths, reduced by 1/4 of the syntonic comma (5.5 Cents), arranged in the circle of fifths, so that eight pure large thirds are produced. The last fifth, the wolf fifth, is too large by 35.5 Cents. The mediant temperament has eight successive, pleasantly sounding major triads (triangles with base line on the interior circle), resp. seven equally such minor triads (triangles with base line on the exterior circle). The remaining triads with their much too large thirds are no longer usable.

**Fig. 2** shows the clear tendency in the organ temperament of WERCKMEISTER to make playable, according to the new needs of baroque music—if only in a limited way—as many triads as possible, and to eliminate the wolf. This comes at the expense of the purity of the large thirds of the good keys in the mediant temperament. There are still triads of different purity.

**Fig. 3** represents LEONHARD EULER's *Genus diatonico-chromaticum*, which has twelve absolutely pure and twelve markedly impure triads. Compare their position with the impure triads in WERCKMEISTER's temperament. However, it is most of all the fifth $D-A$, too small by 21.5 Cents and severely disturbing the sequence of pure triads, which is a hindrance in the practice of music.

*der transposition und anderen vortheilen.*[116] [Your musical work may well be
very interesting: But I doubt that the musicians would accept your temperament; that the general term $2^n \times 3^m \times 5^p$ includes nearly all tones
that are pleasantly perceived is probably to be considered nothing more
than an observation. In music, I don't believe that one ought to reflect
foremost on a most perfect harmony, since with the sense of hearing one
cannot distinguish a comma. Supposing that the geometric progression[115]
gives the tones so accurate that it gives a simple proportion if not accurate
then perceivably accurate, one would have to prefer the latter, because of
transposition and other advantages.[116]]

EULER, however, was sure of his thing and replied: *Meine Theoria ist
durch den Druck schon fast zu Ende gebracht; was ich von dem Termino gener-
ali $2^n \cdot 3^3 \cdot 5^2$ gemeldet, ist nicht nur eine Observation, sondern kommt mit der
neuesten und probatesten Temperatur so genau überein, daß nur der Clavis B ein
wenig different ist ... Wann also in der recipierten Art nur der Ton B in ratione
128:125 tiefer gemacht wird, so komt dieselbe mit der wahren Harmonie überein.
Dadurch wird zugleich das vom* MATTHESON *angeführte inconveniens*[117] *völlig
gehoben, und das intervallum Cs:B in eine sextam majorem verwandelt, welches
sonsten einer septimae minori näher käme. Übrigens ist die Eintheilung secun-
dum progressionem geometricam schon ausgemustert, weil sie alzuviel abweicht
von den wahren Consonantien.*[118] [My theory has almost been completed
by the press; what I said about the general term $2^n \cdot 3^3 \cdot 5^2$ is not only
an observation, but turns out to agree with the newest and most probable temperament so accurately that only the key $B^\flat$ is a little different ...
Thus, when in the recipient manner only the tone $B^\flat$ is lowered by the ratio
128:125, then agreement with the true harmony is achieved. In this way,
at the same time, the inconvenience[117] mentioned by MATTHESON is completely removed and the interval $C^\sharp : B^\flat$ changed into a major sixth, while
otherwise it would be closer to a minor seventh. By the way, the division
according to the geometric progression must be rejected since it deviates
too much from the true consonants.[118]] But DANIEL is stubborn, and it
is a testimony of true scientific spirit in the best sense of the word when
he reports back to Petersburg: *Ich habe mir vorgenommen mit dem hiesigen
Hr.* PFAFF[119] *(der ein vortrefflicher Musikus ist) einen flügel so ich habe auff dero
vorgeschriebene manier stimmen zu lassen: er aber zweiffelt daß solches einen
guten effect tun werde, und müsse man nicht, sagt er, auff die harmonie allein*

*achtung geben, sonderlich wan es de differentiis imperceptibilibus zu thun ist.*[120]
[I took upon me, together with Mr. PFAFF[119] (who is an excellent musician),
to have my piano tuned according to your prescribed manner: However,
he doubts that this will produce a good effect and, he said, one should not
pay attention solely to the harmony, especially when one has to deal with
imperceptible differences.[120]]

Unfortunately, we have no word about the outcome of this experiment — DANIEL BERNOULLI in the course of his correspondence has no
longer addressed this problem of temperament.

In order to be able to clearly compare EULER's temperament with others, with a view to contemporary musical practice, we represent it, after
BEATRICE BOSSHART, with the following fifths-thirds-circle (Fig. 3).[121]

The triangles with the base on the inner circle and the point on the
outer represent the major triads arranged in the usual fifths-circle, the
complementary ones with the base on the outer circle the triads of the
respectively parallel minor keys. Between the individual tones there are
indicated the deviations in Cents from the corresponding pure intervals
(100 Cents correspond to an equally tempered half-tone interval). They are
here positive for fifths that are too small, large thirds that are too large, and
small thirds that are too small. For the equal temperament, they are, in the
above order, 2, 14, and 16 Cents. Werckmeister's temperament (Fig. 2) —
as a paradigm of well-tempering — shows the thirds-characteristic of the
in the 18th century commonly practiced, not equal temperaments. If one
considers the mediant fifths (696.6 Cents) and the equally tempered large
thirds (400 Cents) as standard values of what is still bearable, then Euler's
temperament turns out to be even worse than the mediant temperament
(Fig. 1) with regard to the number of useful major/minor triads. But most
of all, it is the position in the circle of the impure fifth $D - A$ which nar-
rows the freedom of movement by an abrupt transition from pure triads
to strongly impure neighboring triads.

Now (after BEATRICE BOSSHART) EULER's temperament is not meant to be
well-tempering for practical purposes, but to be the simplest *genus musicum*
among other, better, but more complicated ones, with the help of which
structures of contemporary music can, to some extent, be explained, illus-
trated, and possibly even improved in a sort of "musical natural science",

starting from the "true (harmonic) principles" of music under inclusion of a tolerance limit for sensory perception (1 comma).

One could add that EULER's tone $B^\flat$ in his chromatic scale would better be named $A^\#$. Then one could obtain agreement — as must be the case — with the fifths-circle: to $A - E - B - F^\#$ there belong the pure large thirds $C^\# - G^\# - D^\# - A^\#$, and only in this way is the sequence of the pure fifths rigorous, for $A^\#-F$ is no longer a fifth. Therefore, with EULER, there are only two false fifths, namely $D-A$ and $F^\# - C^\#$ (with 680 Cents), and it becomes clear why, with EULER, $B^\flat$-major for example cannot be played: Since there exists not even a tone $B^\flat$, and thus from $C$ also no small seventh.

In summary, one can observe that EULER's tonal system is rather similar to the one of JOHANNES KEPLER[122], but we leave the question aside whether the great man from Basel "from today's perspectives proves himself to be quasi a Goethean[123]", only because he always remained vehemently opposed to the equal temperament. Rather, the idea of a new tonal system — as it appears to us today — originated from a metaphysical-mathematical need, that is, EULER's love of numbers — and of music.[124]

# Farewell from Petersburg

What now moved EULER to depart from Petersburg, even though he obviously felt at home there and very much appreciated the unique opportunities for work in the new city? He himself describes the matter in his short autobiography (p. 4) as follows: *Als hierauf A. 1740 Seine noch glorreich regierende Königl. Majestät [FREDERICK II; EAF] in Preussen zur Regierung kamen, so erhielt ich eine allergnädigste Vocation nach Berlin, welche ich auch, nachdem die glorwürdige Kayserin ANNE verstorben war, und es bey der darauffolgenden Regentschafft ziemlich misslich auszusehen anfieng, ohne einiges Bedencken annahm ...* [Then, in 1740, when His still gloriously reigning Royal Majesty came to power in Prussia, I received a most gracious call to Berlin, which, after the illustrious Empress ANNE had died and it began to look rather dismal in the regency that followed, I accepted without much hesitation ... ]

Surely, the increasing gravity of the internal situation in Russia was a weighty reason for EULER to leave the country. After the death of ANNA IVANOVNA (end of 1740), the accession to the throne of the infant IVAN VI, the "MÜNNICH-OSTERMANN-Coup", and the palace revolt of 1741, which was to put ELISAVETTA PETROVNA for twenty years on the Russian throne[125], the situation in the course of the "russification" by PETER's daughter certainly did not look rosy for "foreigners", though EULER, in view of his services for Russia in general, and the Academy in particular, would probably not have had to fear anything serious – at least that is what E. WINTER[126] thinks, but EULER still had at least three other reasons for his transfer. The geography, which was fatal to him, was already mentioned above (p. 41), but in addition to this, there was something quite different. As we can gather from a very intimate letter of EULER to MÜLLER of July 8/19, 1763, which he wrote in the context of negotiations for the call back to Petersburg, it was especially his wife KATHARINA who urged him to accept the call to Berlin, because she was apprehensive of the fires that frequently broke out in Petersburg; this was also the reason why the escape luggage was permanently kept in readiness in EULER's (wooden) house. But let us hear EULER himself:

*Ich muß aber dabey auch mit auf meine Familie sehen, von welcher ich mich unmöglich trennen kan und welche sich noch vor einigen Umständen förchtet, die vor 22 Jahren das meiste zu meiner Abreise beygetragen, nunmehro aber, wie ich vermuthe, solches aber doch meiner Frau noch nicht bereden kan, sich gänzlich geändert haben müssen. Meine Frau kan nehmlich noch nicht ohne Schrecken an die damaligen Zeiten gedencken, da man aus Furcht vor Brand alle seine Habseeligkeiten beständig eingepackt halten muste, um solche bey entstehender Gefahr desto füglicher zu retten, wobey man aber sich mit der traurigen Vorstellung immer quälen muste, wenn man allzuplötzlich überfallen würde, alles das Seinige auf einmal einzubüßen, wie damals öfters vielen Leuten begegnet, welche sogar diejenige Küste, darinn ihre besten Sachen verwahret waren, haben im Stich lassen müssen. Eur. Hochedelgeb. werden leicht erkennen, daß diese beständige ängstliche Furcht alle Vortheile, die man sonst genießen könnte, zernichten müsse. Ich glaube zwar aus der jetzigen Verfassung schließen zu können, daß nunmehro bessere anstalten gegen die Feuersgefahr vorgekehret worden, doch aber, um auch meine Familie darüber zu beruhigen, so wünschte ich sehr, eine gründliche Bestätigung zu erhalten.*[127] [I must, however, in this matter also keep my family in mind, from whom I cannot possibly separate, and who still is afraid of some circumstances 22 years ago that contributed mostly to my departure, but now, I suspect, these must have changed completely, though I cannot convince my wife thereof. My wife, namely, can still not think of those times without horror when, for fear of fire, one had to keep all one's belongings permanently packed in order to be able to more readily salvage them when danger arose, whereby however one continuously had to torment oneself with the sad prospect of losing one's possessions all at once if the attack came too suddenly, which then often happened to many people, who had to abandon even the coffer in which they had stored their best things. Your Honorable will readily understand that this lingering fear necessarily shatters all the advantages that could otherwise be enjoyed. I believe, though, that from the current condition one can conclude that nowadays better preventive measures have been instituted against danger of fire, but, however, in order to also calm down my wife about this, I would greatly appreciate to receive a definitive confirmation.[127]]

That such fears were not unfounded was to be later confirmed in a horrible way: In the terrible fire of May 1771 in Petersburg, which over 550 houses fell victim to — among them also EULER's — the almost blind mathematician would have been burnt alive if not PETER GRIMM, a brave craftsman from Basel, had rescued him from the burning house at the risk of his own life.[128] A blessing in disguise was the fact that most of EULER's manuscripts, which were ready to go to press, were safe, thanks to the prudence of ORLOV[129].

Another reason, though less weighty, which led EULER to leave Petersburg may have been the burden of forced quartering of officers and soldiers; this, in view of the great shortage of dwellings in Petersburg, appears to have been unavoidable and quite common.

Yet, the departure was not made easy for EULER. Even though he was a free man according to the wording of the contract with the Petersburg Academy, the authoritarian chancellor SCHUMACHER pressed against EULER's discharge with all his might, repeatedly claiming that EULER's presence "is a necessity for the Academy". It required the full influence of the president of the Academy, v. BREVERN, of GOLDBACH, and of — the then still powerful — Privy Councilor HEINRICH JOHANN OSTERMANN, as also the emphatic intervention of the Prussian envoy in Petersburg, AXEL VON MARDEFELD, in order to have the mathematician, taken ill by this agitation, released with the official reason that EULER's state of health made this necessary.

# 3
## The Berlin period 1741–1766

## New beginning in FREDERICK's Prussia

On June 19, 1741, EULER set out on the journey from St. Petersburg to Berlin, with his wife KATHARINA, the first-born JOHANN ALBRECHT, the one-year-old toddler KARL and — probably — the brother JOHANN HEINRICH[131] — in the luggage the calling document of FREDERICK II, which had been delivered to him on February 15, 1741, by VON MARDEFELD, and in which he was assured a yearly salary of 1,600 Taler along with reimbursement for travel expenses in the amount of 500 Taler.[132] After a three-week's journey on sea, the party arrived in Stettin, where EULER was greeted most cordially by various honorabilities and visited a disputation at the "Gymnasium". On July 22, the journey continued by coach and carriage toward Berlin, where the EULER family happily arrived at last on July 25.[133]

Naturally, EULER had informed his parents in Basel early about his plans. From the first reaction — authenticated still today — namely the one of DANIEL BERNOULLI, we learn that EULER at first intended to go to Berlin alone, and that the king FREDERICK II had invited to Berlin JOHANN I BERNOULLI and also the latter's sons DANIEL and JOHANN II: *Zu der herrlichen Berliner vocation hingegen gratuliere ich von hertzen; ich erfrewe mich zum voraus, daß ich nocheinmahl die ehr haben solle Ew. HEdgb. zu sehen; doch bedauere ich daß Sie nicht Dero samtliche famille mit sich führen wollen, wie mir Dero Hr. Vatter meldet. Ich werde vielleicht dessen ohngeacht noch die ehr haben Sie zu sehen, dan ich im sinn habe mit der zeit eine reyss nach Berlin zu thun. Ihre May[estät] haben meinen Vatter, meinen Bruder und mich auch invitieren lassen. Mein Vatter hat sich völlig excusiert; ich habe mich auch noch nicht resolvieren können. Mein bruder aber möchte wohl die vocation annehmen. Es ist underdessen zu beförchten, daß der krieg das ganze project, wo nicht völlig stöhre doch aufhalte.*[134] [For the splendid call to Berlin, however, I congratulate you sincerely; I look forward with joy to have once again the honor of seeing you; yet I regret that you will not come with your whole family, as your father told me. In spite of that, I will probably still have the honor of seeing you, since it is my intention to undertake, in due time, a journey to Berlin. His Majesty has invited also my father, my brother, and myself. My father excused himself entirely; I myself also have not yet been able to decide. My brother, however, may well want to accept the call. In the meantime, I fear that the war may, if not completely upset, then at least delay, the whole project.[134]]

EULER's old teacher in Basel also reacted similarly in his letter of February 18, 1741, which because of the uncertainty of EULER's address he actually held back and mailed it to Berlin only on September 1 — provided with an explanatory P. S. : "I have heard with joy . . . that you have been invited on behalf of the Prussian king to organize the new Academy in Berlin, and that you already accepted this call, for the honor of which I sincerely congratulate you. May God assist you in this endeavor and accompany you on this journey, which to my knowledge you in fact have to set out on soon in June already. May I ask you to write to me what yearly salary you have been promised? I, too, and my two sons have, at the direction of the king, received letters of invitation, but I am too old, and my health is too shaky, to lend, as I would like, an ear to this so honorable and tempting enticement. Were I twenty years younger, good gracious!, I wouldn't hesitate for one moment; I am sick and tired of it all in the homeland. What decisions my sons will make, I don't know yet; I believe, they will wait for more precise information about the conditions of employment, and that will be forthcoming — if I may be allowed to speculate — as soon as the current Silesian campaign comes to an end[135]. Once you are in Berlin, we will have you much closer, and that lets me hope that some time or other you will make a trip to hearth and home to greet your parents, which would give me the opportunity of seeing you, which I most ardently desire before I will die."[136]

The short letter of September 4, 1741, from the camp of Reichenbach[137], with which FREDERICK II officially welcomed EULER, consists essentially of the confirmation of EULER's salary and closes with the remark that were he, EULER, still in need of anything, he only has to await the return of the king to Berlin. As a matter of fact, FREDERICK already at the beginning of his assuming power (1740) devised big plans for the establishment of a new Academy in Berlin, which next to the existing Academies in London,

FREDERICK II. Oil painting by A. GRAFF

Paris, and Petersburg, would make a good showing, and about this, the new king of Prussia must have advised also EULER — through VON MARDEFELD — already in Petersburg. In Berlin, EULER, above all, wanted to work, and his head was full of plans and ideas. Yet, at the time of EULER's arrival, FREDERICK II was engaged in a serious conflict with the Danubian monarchy, brought about by himself, and as is well known, "the Muses are silent in the noise of arms". Thus, the creation of the Berlin Academy proved to be a forceps delivery. On the basis of the "Kurfürstlich-Brandenburgischen Sozietät der Wissenschaften", founded in July of 1700 by the Elector FREDERICK III, whose actual creator and first president was LEIBNIZ, there evolved the first Academy — distinctly national in character — under the name "Königlich Preußische Sozietät der Wissenschaften"[138]. It suffered a noticeable decline under the rule of FREDERICK WILHELM I[139], the "military king", who, as is known, expressed his disdain for the Academy and scholars in general also by the fact that he appointed in 1731 and 1732 a "Royal Joker" (that is, a court jester) as president resp. vice president of the Academy. About the most important scholar of his time, LEIBNIZ, he is known to have said: "The chap is no good for anything, not even to stand guard."[140]

Subsequently, in 1744, the "Sozietät" was united by FREDERICK II with the "Nouvelle Société Littéraire" founded in 1743 — to which we will return later — to become the "Königliche Akademie der Wissenschaften"[141], which thereafter, in 1746, received the name "Académie Royale des Sciences et Belles-Lettres", with the new statute under the presidency of MAU-PERTUIS, and became in all respects thoroughly "frenchised".

## EULER's first activities in Berlin

EULER's first duty at the Berlin Academy, more precisely, the "Society" — its predecessor — , was the compilation of the seventh volume of the *Miscellanea Berolinensia*, which had been started already before him, and about which EULER on December 15, 1742, could report to GOLDBACH: *Mit dem neuen tomo Miscellaneorum Berolinensium ist man schon ziemlich weit gekommen, worin fast die ganze Classis Mathematica von mir kommt.*[143] [The new volume of *Miscellaneorum Berolinensium* has progressed quite well, wherein almost all of the Mathematical Class is by myself.[143]] Indeed, the protocol of the first session of the Mathematical Class, in which EULER was introduced as a new member, shows that "various members and especially the famous mathematics professor Mister EULER, whom His Majesty has called from the Petersburg Academy, have submitted, or are about to submit, their contribution"[144]. EULER in this session read seven of his works which he had written since his arrival in Berlin and now made available to the *Miscellanea Berolinensia* to be printed. They consist of memoirs 1. on the determination of the orbit of the comet[145] observed in March of 1742, 2. on theorems about the reduction of certain integral formulae to the quadrature of the circle[146], 3. on the theory of definite integrals[147], 4. on the summation of reciprocal series formed with the powers of natural numbers[148], 5. on the integration of differential equations of higher order[149], 6. on certain properties of conical sections which correspond to a family of other curves[150], and finally 7. on the solution of a special linear differential equation[151].

The first five of these papers were included in Volume 7 of the "Miscellanea Berolinensia", while the last two, for reasons of space, were held back for a supplementary volume, as we are told in the protocol[152].

But also with regard to private affairs, EULER seems to have established himself well, even though in Berlin he had to wait for an entire quarter of a year until his first salary and travel expenses were paid, and had to live on credit. At the end of 1742, he wrote to his friend GOLDBACH: *Der H. Brigadier* BAUDAN *ist hier noch außer Dienst, und hat sich mit einer Mlle.* MIRABEL *verheurated, deren Vermögen ungefähr in 4,000 Rth. bestehet, darunter ein artiges Haus begriffen war, welches ich für 2,000 Rth. gekauft und dazu von Ihro Königl. Majestät das Privilegium eines Freihauses erhalten habe. Dasselbe liegt zwischen der Friedrichs- und Dorotheenstadt, nahe bei dem Ort, wo Ihro Majestät, der König, das neue Schloß und die Akademie zu bauen beschlossen hat, daß also die Situation nicht erwünschter sein könnte.* [Mr. Brigadier BAUDAN is here still off duty, and married a certain Mlle. MIRABEL, whose fortune is approximately 4,000 Rth., which includes a nice house, which I bought for 2,000 Rth. and, in addition, received from His Royal Majesty the privilege of a free house. The latter lies between the Friedrichs- and Dorotheenstadt, near the place where His Majesty, the King, has decided to build the new castle and the Academy, so that the situation could not be more desirable.] It is true, however, that EULER could move into this house, which *Ihre Königl. Majestät auf mein Alleruntertänigstes Ansuchen auf ewig von aller Einquartierung loszusprechen allergnädigst geruhet*[153] [His Royal Majesty, upon my most humble petition, has most graciously agreed to keep forever free of any quartering whatsoever[153]], only on Michaelis 1743 — thus almost a year after the purchase — since the building required various repairs in the amount *von etlichen hundert Rth.* [of several hundred Rth. ]. Even though, in the meantime, EULER's family, with the daughter KATHARINA HELENE[154] and the son CHRISTOPH[155], had grown to six persons, the house offered room in abundance, so that the — still very young — GREGORIY TEPLOV and the future president of the Petersburg Academy, KIRILL RASUMOVSKI, could *vergnügt beisammen wohnen*[156] [merrily live together[156]] as pensioners in EULER's household. This type of hospitality EULER has granted throughout his stay in Berlin, as we can learn from various correspondences.

EULER made excellent use of the war years until the actual opening of the Berlin Academy in January of 1746. Apart from the seven already mentioned, and twelve additional, memoirs, one of which earned him the prize of the Paris Academy, the indefatigable EULER wrote about 200 letters, some very extensive, and five books, which today are counted among his principal works[157]. Of these memoirs, one is especially fascinating inasmuch as it is evidently tailored to nonmathematicians; quite possibly, EU-

LER could have written it for his new employer, King FREDERICK II, were it not composed in Latin. The memoir *On the usefulness of higher mathematics* is so characteristic of EULER's clear and easily comprehensible style, that we quote here from its introductory section:

*Today nobody doubts the great usefulness of mathematics, because for many sciences and arts, which we make daily use of, it is indispensable. This praise now, however, is usually rendered upon the lower mathematics, upon its elements, as it were, while for the kind of mathematics which, rightly so, is called higher, any practical importance thereof is disputed. It were a spider's web, one argues, which, owing to its extraordinary delicateness, cannot be used. All of mathematics, however, is concerned with the search for unknown quantities. To this end it shows us the methods or the paths, as it were, which lead to the truth; it finds the most secret truths and puts them in the right light. In this way, on the one hand, it sharpens our mental power, but also, on the other hand, enriches our knowledge. Both are goals which surely are worthy of the greatest effort. The truth in itself is a jewel; since several truths, linked together, yield higher relationships, each one is useful, even if this at first is not evident. One sometimes also objects that the higher mathematics lets itself sink too deeply into the exploration of the truth. This is more a praise than a criticism.*

*But let us not dwell on these abstract preferences, since, after all, we can easily prove that higher analysis does not merit the designation of a useful science with less right than elementary mathematics, indeed even a much wider field of application opens up to it. Even in those sciences in which elementary mathematics at first seemed to be generally sufficient, a further development of higher mathematics is necessary to a degree which it has by far not yet attained. Therefore, in this memoir I want to show that the usefulness one attributes to elementary mathematics surely does not disappear in higher mathematics, but to the contrary, continuously increases, the higher one climbs in this science; indeed, that mathematics has not even developed as far as even the most common applications actually would require.*[158]

Thereafter, EULER discusses with words equally concise the true usefulness of higher mathematics in such fields as mechanics, hydraulics, astronomy including optics, artillery (with greetings from FREDERICK II!), physics, and physiology, and he concludes with the words: *I believe to have reached the goal I aimed for, namely, to clearly accentuate the great usefulness of higher analysis. Other reasons in great number could corroborate my proof. I could demonstrate that analysis sharpens our mental power and thus prepares for establishing the truth. But the enemies of mathematics could find here material for disputes. The proofs I gave are irrefutible, and with this I feel satisfied.*

# Other major works

## Calculus of variations*

Following various problems posed, and ideas expressed, by JAKOB and JO-HANN BERNOULLI, EULER already early — first in a long (largely unsuccessful) memoir[159], then in his celebrated *Methodus*[160] — formulated the main problems of the "calculus of variations" and developed general methods for their solution.[161] This special discipline — initiated in rudimentary form by the brothers BERNOULLI, conceptualized and systematized for the first time by EULER — deals with extremal problems of the most general kind. While in the differential calculus one deals, among other things, with the problem of determining those values of one or several variables for which a given function attains a maximum or minimum, the problems in the calculus of variation are characterized by having to determine one or several unknown functions in such a way that a given "definite integral", which depends on these functions, attains extremal values. In general, the solution of the problem leads to "EULER's differential equation". It is probably rarely the case that a book has been written whose content is so accurately expressed in its title as it is in EULER's *Methodus*, which appeared in 1744. The method is primarily geometric, but it allowed EULER to discover the so-called isoparametric rule, which already contains in itself the germ of a generalization and formal condensation — amounting to a new formulation — which the calculus of variation has subsequently, at the beginning of the sixties, undergone in the hands of the congenial LAGRANGE.[162] EULER immediately subscribed to this significant step, which outdistanced even himself; he abandoned his old method and developed the principles of the new method[163] on the basis of the extremely precise LAGRANGian algorithm, illustrating them — typical for EULER — with numerous applications. "It always means progress if one succeeds in adding to EULER's

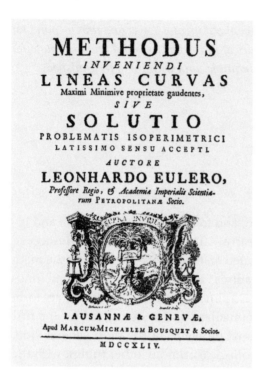

Frontispiece of LEONHARD EULER's "Calculus of variations", Lausanne and Geneva 1744

examples a truly new one" remarked one of the most important mathematicians of the 19th century, C. G. J. JACOBI, who had a very high esteem of EULER's overall work and also engaged himself vigorously with an edition of his works.[164] JACOBI's remark relates to the fact that EULER in his *Methodus*, with its 66-page *Additamenta*, has carried out all the calculations necessary in many — almost all — applied examples relevant to the research areas of his time (elasticity, bending of a beam, ballistics, etc.).

## Ballistics

In this subsection, we are talking about artillery. One should beware being tempted to view EULER's occupation with this "military subject" from a narrow, pacifist angle; the determination of the "ballistic curve", that is, the trajectory of a (spherical) projectile with air resistance taken into account, was since the middle of the 16th century one of the most demanding subjects dealt with by the most important mathematicians like TARTAGLIA, BENEDETTI, GALILEI, NEWTON, HUYGENS, JOHANN and DANIEL BERNOULLI, and plays a considerable role in the history of mathematics and physics.[165]

NIKOLAUS FUSS[166] tells us: "The king [FREDERICK II] had requested Mister EULER's opinion about the best work in this area. From ROBINS, who a few years earlier had rudely attacked EULER's Mechanics (of 1736), which he did not understand, there appeared new principles of artillery in English, which Mister EULER praised to the king, at the same time pledging to translate the work and to add notes and explanations. These explanations contain a complete theory of the motion of thrown bodies, and nothing has appeared in the last 38 years that comes close to what Mister EULER at that time had done in this difficult part of mechanics. Also, the value of this wonderful work was widely recognized. An enlightened statesman, the French maritime and finance minister TURGOT, had it translated [1783] into French and introduced in the schools of artillery, and almost at the same time [1777] an English translation appeared in the greatest typographical splendor that English printers were able to muster. While Mister EULER, whenever appropriate, did justice to Mister ROBINS, he corrected with rare modesty the latter's error against the theory, and the only revenge he took against his adversary because of the old injustice consists in the fact that he made the latter's work so famous as, without him, it would never have become."[167]

Let us briefly consider the facts of this story, which after all, FUSS could relate almost firsthand.[168] Subsequent to EULER's *Mechanica*[169], BENJAMIN ROBINS[170] — and this must be viewed in the context of the "kinetic energy controversy of JOHANNN BERNOULLI with the Englishmen TAYLOR, KEILL, JURIN, and MACLAURIN — went so far as to make the following provocative statement: "I don't have the intention to accuse the author [EULER] about his errors of haste or negligence, but I consider them solely a consequence of the inaccuracies in the formation of concepts, to which the differential calculus can mislead its admirers ... At the beginning of the third chapter, which deals with rectilinear motion, Mister EULER presents GALILEI's theory of falling bodies, which in itself is not a difficult topic, but is here mixed with differential calculus to such an extent that one better looks up this subject in a place where it is written up by others in a simpler manner."[171] EULER did not react in any way to these reproaches, but when in 1742

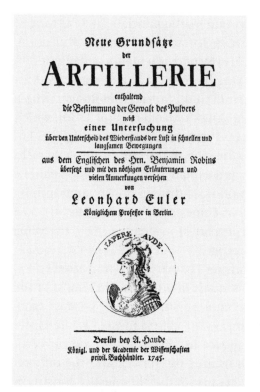

Neue Grundsätze
der

# ARTILLERIE

enthaltend
die Bestimmung der Gewalt des Pulvers
nebst
einer Untersuchung
über den Unterscheid des Wiederstands der Luft in schnellen und
langsamen Bewegungen

aus dem Englischen des Hrn. Benjamin Robins
übersetzt und mit den nöthigen Erläuterungen und
vielen Anmerkungen versehen
von

Leonhard Euler
Königlichem Professor in Berlin.

Berlin bey A. Haude
Königl. und der Academie der Wissenschaften
privil. Buchhändler. 1745.

Frontispiece of LEONHARD EULER's "Artillery", Berlin 1745

ROBINS's principal work[172] appeared, he recognized it immediately as a decisive breakthrough in the field of ballistics, translated the booklet into German — surely a sort of welcome salute to FREDERICK II — and provided it with "vielen Anmerkungen" [many notes], which in themselves made up about four times the size of the English original. They constitute the first treatment of inner, outer, and target ballistics, systematically using the LEIBNIZ-BERNOULLI infinitesimal calculus, and contain, among other things, the trajectory of the oblique shot, the "ballistic line" (in whose determination even NEWTON still did not succeed) in parametric form and the resulting power series approximation.[173]

EULER's *Neue Grundsätze der Artillerie*[174] [New principles of gunnery[174]] — one of the few books, incidentally, which EULER published in the German language — appeared in 1745 and was translated 1777 into English, and 1783 into French[175], as we already could gather from the report of FUSS, and thereby found entry into all schools of artillery.

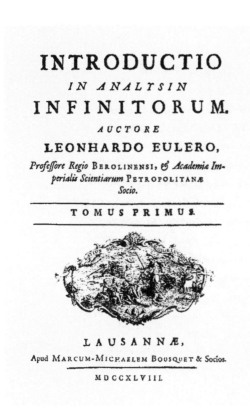

**INTRODUCTIO**
*IN ANALYSIN*
**INFINITORUM.**
*AUCTORE*
**LEONHARDO EULERO,**
*Profeffore Regio* BEROLINENSI, *& Academia Im-*
*perialis Scientiarum* PETROPOLITANÆ
*Socio.*

TOMUS PRIMUS.

LAUSANNÆ,
Apud MARCUM-MICHAELEM BOUSQUET & Socios.

MDCCXLVIII.

Frontispiece of EULER's "Introduction into the analysis of the infinite",
Lausanne 1748

As to EULER's truly noble form of revenge against ROBINS, there was someone who certainly did not at all agree with it: the old "lion" in Basel. On September 23, 1745, JOHANN BERNOULLI writes to his master pupil in Berlin: "The first of these books [E. 77] I have already read through almost completely, but admittedly in such a way that I assumed the correctness of your calculation and did not recheck the calculations for myself; because most of them appeared to me to be too complicated, so that in view of my bad state of health I did not dare to take them on. But do you believe that ROBINS, an Englishman, will be able to understand your German notes to this treatise published on the same subject? I wonder about your leniency and your courtesy toward ROBINS, who after all expressed himself only mockingly about yourself, about me and all nonEnglishmen[176]. He calls you — as I hear — a computing machine, as if you wouldn't function other than a machine driven by a weight. Me, however, he jostles in an extremely

Caput. 2.

De Calculo Differentiali in genere.

§.1. Ex iis, quæ de calculo differentiarum finitarum prolata sunt, facile erit intelligere, quid sit calculus differentialis seu differentiarum infinite parvarum. Ibi enim inveniebatur incrementum functionis, quantitatibus radicalibus augmenta finitæ magnitudinis accipientibus, seu potius augmenta quæcunq; Hic vero hæc augmenta infinite parva ponuntur, quaritur q ex quantitatum functionem componentium incrementis infinite parvis, incrementum totius functionis. Docet igitur Calculus differentialis functionis cujuscunq incrementum invenire ex datis quantitatum eam ingredientium incrementis infinite parvis. Hæc incrementa infinite parva vocantur differentialia et quantitatem differentiare significat ejus differentiale invenire.

§.2. Cum in capite præcedente tractatum sit de inveniendo incremento functionis, quantitatibus radicalibus incrementa quæcunq accipientibus, hic vero agatur de inveniendo incremento functionis, si incrementa quantitatum eam componentium fuerint infinite parva: Perspicuum est Calculum differentialem, ejus quem ante exposui, calculi esse casum specialem: nam quod ibi quantum erat a quantum, hic ponitur infinite parvum. Quæ igitur ante tradita sunt, et hic valent, sed præterea hoc in casu alia adjungi debent, ex infinite parvis deducenda.

First page of the second chapter of EULER's manuscript for the "Differential calculus"

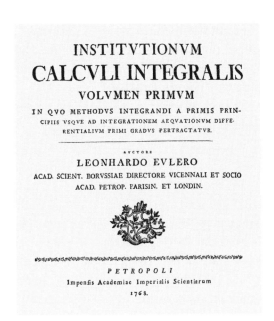

INSTITVTIONVM
# CALCVLI INTEGRALIS
## VOLVMEN PRIMVM
IN QVO METHODVS INTEGRANDI A PRIMIS PRIN-
CIPIIS VSQVE AD INTEGRATIONEM AEQVATIONVM DIFFE-
RENTIALIVM PRIMI GRADVS PERTRACTATVR.

AVCTORE
### LEONHARDO EVLERO
ACAD. SCIENT. BORVSSIAE DIRECTORE VICENNALI ET SOCIO
ACAD. PETROP. PARISIN. ET LONDIN.

*PETROPOLI*
Impenſis Academiae Imperialis Scientiarum
1768.

Frontispiece of LEONHARD EULER's "Integral calculus", Petersburg 1768

ludicrous way, because in my French memoir 'Sur le mouvement'[177] I advocated the kinetic energy. However to you ... I thank for the honorable praise with which you had the kindness of showering me in this work several times."[178]

Then there follows a tirade — pages long — about BERNOULLI's painful experiences with KEILL and TAYLOR in connection with the struggle about the ballistic curve.

Whether and how EULER reacted to BERNOULLI's faint reproach, we do not know in the absence of the response letter. Probably, he let the matter rest, since quarrels between scholars interested him far less than his own research work: At the same time as his *Calculus of variations*, there in fact appeared his book on the theory of planets and comets[179], which underwent a translation into German by JOHANN VON PACCASSI (Vienna 1781), and to which there immediately followed (anonymously) two supplements.[180] Furthermore, into the same period falls EULER's *New theory of light and colors*[181], wherein he developed his entire program for the optics, from which, also later, he hardly was to significantly deviate anymore. "The work addresses all phenomena then known; on the one hand, those of light: the act of seeing in general, the formation and propagation of light, effects of impulses and light rays, reflection and refraction; on the other hand, optical properties of matter: luminous, reflecting, refracting the light, and opaque bodies ... Above all, EULER grapples with NEWTON's corpuscular optics, which he now contrasts with his own wave theory."[182]

We will still go later into EULER's optical writings, which, with seven volumes, account for about ten percent of the printed overall works.

## The *Introductio* of 1748

EULER's two-volume work *Introduction into the analysis of the infinite*, the *Introductio*[183], as it is known for short, is the starting signal for a "great trilogy", which — with the likewise two-part *Differential calculus*[184], already begun years before the *Introductio*, and the substantially later four-volume *Integral calculus*[185] — represents a magnificent synopsis of the most important mathematical discoveries in analysis until, and beyond, the middle of the 18th century, and which for generations has become the model and foundation of the textbook literature.

EULER's *Introductio*, the first part of which is dedicated to the theory of the so-called elementary functions, without any use being made of infinitesimal calculus, actually represents the beginning of function theory, a principal branch of modern mathematics, which subsequently was to undergo a huge, brisk development — a development whose end even today cannot yet be foreseen. Of central importance in the *Introductio* is the elaboration of the analytical concept of a function, as well as EULER's clear observation that mathematical analysis is to be regarded as the *science of functions*, and nothing short of a mathematical-historical caesura is EULER's formation of the concept of *complex functions*[186].

EULER expressed his own opinion about this book already in his letter of July 4, 1744, to GOLDBACH: *Ich habe inzwischen ein neues Werk dahin* [nach Lausanne] *geschickt unter dem Titel "Introductio ad Analysin infinitorum", worin ich sowohl den partem sublimiorem der Algeber* [!] *als der Geometrie abgehandelt und eine große Menge schwerer problematum ohne den calculum infinitesimalem resolviert, wovon fast nichts anderswo anzutreffen. Nachdem ich mir einen Plan von einem vollständigen Traktat über die Analysis infinitorum formiert hatte, so habe ich bemerkt, daß sehr viele Sachen, welche dazu eigentlich nicht gehören und nirgend abgehandelt gefunden werden, vorhergehen müßten, und aus denselben ist dieses Werk als Prodromus ad Analysin infinitorum entstanden.*[187] [In the meantime I sent there [to Lausanne] a new work entitled "Introductio ad Analysin infinitorum", where I treated the higher parts of both the algebra and geometry and solve a large number of difficult problems without the infinitesimal calculus, of which almost nothing can be found elsewhere. After I had worked out a plan for a complete treatise on infinitesimal analysis, I noticed that a great many things, which really do not belong here, and are not treated anywhere, must be mentioned beforehand, and from these, the present work ensued as a precursor to infinitesimal analysis.[187]] The editors of the first volume emphasize that the work "still today deserves to be not only read, but studied with devotion. No mathematician will put it aside without immense benefit", and that the *Introductio* "marks the beginning of a new epoch and that this work has become influential for the whole development of the mathematical sciences by virtue of not only its content, but also its language"[188]. ANDREAS SPEISER goes still one step further and comments (1945) on EULER's masterpiece with the following emphatic words: "It may well be the counterpart of the usual schoolbooks and could be used as a model for the complete overhaul of mathematical instruction in the secondary schools ... But much water will still flow down the Rhine until schools finally realize that mathematics can be an art and students can understand EULER equally well as PLATO or GOETHE."[189] Well, this critical amount of Rhine water still today has not yet been reached, and SPEISER's wishful dreams seem to become more and more removed from any possible realization because of the present (and future) overemphasis in schools on computer science.

## The philosopher[190]

EULER's philosophical legacy are the *Letters to a German princess*[191], which were written in the French language 1760–1762 to the youthful Margravine SOPHIE CHARLOTTE FREDERIKE VON BRANDENBURG-SCHWEDT[192] and her younger sister LOUISE on behalf of their father. These *Letters* were published in Petersburg in 1768 and 1772 and appeared in three volumes. They

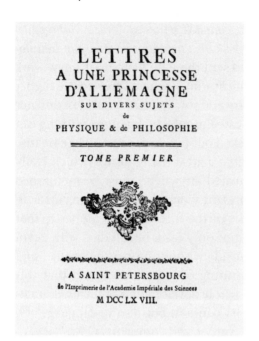

LETTRES
A UNE PRINCESSE
D'ALLEMAGNE
SUR DIVERS SUJETS
de
PHYSIQUE & de PHILOSOPHIE

TOME PREMIER

A SAINT PETERSBOURG
de l'Imprimerie de l'Academie Impériale des Sciences
M DCC LX VIII.

Frontispiece of LEONHARD EULER's "Philosophical letters", Petersburg 1768

became an immediate hit: They were quickly translated into all major languages and for a long time were the most widely distributed synopsis of popular scientific and philosophical culture.[193] The 234 letters comprise music theory, philosophy, mechanics, optics, astronomy, theology, and ethics almost in equal parts, and culminate in the attempt to refute absolute idealism (solipsism) in the sense of BERKELEY[194] and the general ideas of HUME (letter 117), as well as in a large-scale attack against the then widely held monadism in the vein of WOLFF (letters 122–132), which unfortunately is identified much too often with that of LEIBNIZ.

EULER's position in the history of philosophy is controversial even to our days, and it is perhaps not all that accidental that the most extreme positions in this matter are taken, of all people, by two mathematicians of Basel, namely OTTO SPIESS (1878–1966) and ANDREAS SPEISER (1885–1970). The former strongly concurs with the assessment — also endorsed by HARNACK — of EULER's most prominent contemporaries, according to which "it is incredible that such a great genius in geometry and analysis finds himself, in metaphysics, beneath the smallest school boy, not to speak of so much banality and absurdity. It may well be said: the Gods did not grant everything to one and the same"[195], and EULER in this connection had to be told by his best friend DANIEL BERNOULLI: *"Sie sollten sich nicht über dergleichen Materien einlassen, denn von Ihnen erwartet man nichts als sublime Sachen, und es ist nicht möglich, in jenen zu excellieren".*[196] [You should not get involved in matters of this kind, because from you one expects nothing but sublime things, and it is not possible to excel in the former.[196]]

OTTO SPIESS sees the deeper reason of EULER's position against the WOLFFians not in an intellectually, but emotionally motivated context: The vehement attacks of the orthodox and religious mathematician, surrounded by so many "free-thinkers" à la VOLTAIRE, D'ARGENS, and DE LA METTRIE, against the monadism which threatened to subjugate to reason even moral categorical questions, are to be understood as apology of Christianity; EULER indeed re-emphasized this in his (1747 anonymously published) pamphlet *Rettung der göttlichen Offenbarung gegen die Einwürfe der Freygeister*[197] [Salvation of the divine revelation against the objections of the free-thinkers[197]], following the notorious academic prize question of 1747 about monadism.[198]

ANDREAS SPEISER, on the other hand, attributes to EULER nothing short of the beginning of modern philosophy, wanting to see KANT in direct relationship with EULER's *Reflections on space and time*[199] of 1748 and the *Letters*.[200] Referring to RIEHL and TIMERDING[201], most philosophers are nevertheless in agreement with the fact that a direct influence of EULER on KANT cannot be denied at least in two respects: First with regard to EULER's thesis that space and time are no abstractions from the world of the senses, upon which KANT substantially based his *transcendental aesthetics,* and secondly with regard to the thesis, postulated by EULER in 1750 and taken over by KANT, about the impenetrability of matter, with which EULER tried to explain certain forces in nature as "short-range forces" (Cartesian) in contrast to the hypothesis of "long-range forces" (Newton). On this point, incidentally, EULER in turn may have been influenced by BAUMGARTEN's *Metaphysics* (1739), which, as is documented, made a great impression on KANT.[202] Finally, EULER secretly perceives in his conviction the main weakness of his own philosophy: There remains the dilemma between the infinite divisibility of space and the finite divisibility of matter. But let us hear the philosopher in his own words: *"Die Piece de Monadibus, welche bei uns das Praemium erhalten*[203]*, hat meine völlige Approbation, als welcher ich auch* *mein votum gegeben. In derselben ist das ganze Lehrgebäude der Monaden völlig zerstöret. Wir haben über diese Materie 30 Piecen bekommen, von welchen noch 6 der besten, sowohl pro als contra monades, gedruckt werden. In denselben ist beiderseits zum wenigsten die Sach so deutlich ausgeführt, daß die bisherigen Klagen, als wann man einander nicht recht verstanden, ins künftige gänzlich aufhören werden. Die ganze Sach beruhet auf der Auswicklung dieses Raisonnements: die Körper sind divisibel; diese Divisibilität gehet entweder immer ohne Ende weiter fort oder nur bis zu einem gewissen Ziel, da man auf solche Dinge kommt, welche nicht weiter teilbar sind. Im letztern Fall hat man die Monaden; im erstern, die divisibilitatem in infinitum, welche zwei Sätze einander so e diametro entgegengesetzt sind, daß davon notwendig der eine wahr, der andere aber falsch sein muß. Alle argumenta pro monadibus gründen sich hauptsächlich auf scheinbare Absurditäten, womit die divisibilitas in infinitum verknüpfet sein soll. Da man sich aber meistenteils von diesem infinito verkehrte Ideen gemacht, so fallen auch dieselben Absurditäten weg. Die Meinung der Monaden zerteilet sich wieder in zwei Parteien, wovon die eine den Monaden alle Ausdehnung gänzlich abspricht, die andere aber dieselben für ausgedehnt hält, jedoch ohne daß sie partes hätten und folglich divisibel wären, welche letztere Meinung meines Erachtens am leichtesten zu refutieren ist. Diejenigen, welche monades magnitudinis expertes statuieren, müssen endlich zugeben, daß auch aus der Zusammensetzung derselben kein extensum entstehen könnte, und sind dahero genötiget, sowohl die Extension als die Körper selbst für bloße Phaenomena und Phantasmata zu halten, ungeacht sie bei dem Anfang ihres ratiocinii die Körper als reell angesehen; dergestalt, da, wann der Schluß wahr wäre, die praemissae notwendig falsch sein müßten.*[204] [The work on monads, which received here the prize[203], has my full approval and hence received my vote. In it, the whole doctrinal building of the monads is completely destroyed. On this topic we have received 30 pieces of work, of which the six best, some for and some against monads, are printed. In these, the matter on either side has at least been explained so clearly that the previous complaints of not having correctly understood one another will have to be dropped entirely in the future. The whole matter comes down to the following argument: Matter is divisible; this divisibility either continuously goes on without end, or continues up to a point where things are encountered which can no longer be divided further. In the latter case, one has the monads, in the former infinite divisibility, two positions which are so diametrically opposed to each other that one necessarily has to be true and the other false. All arguments in favor of monads are mainly based on seeming absurdities which infinite divisibility is supposed to be entangled with. Since, however, in most cases one has had wrong ideas about this infinity, these absurdities then in fact disappear. The supporters of monads divide further into two parties, of which one denies to monads any expansion whatsoever, while the other admits monads to have expansion, but without possessing parts and thus without being divisible, which latter opinion in my view can be most easily refuted. Those who do not attribute any spatial extension to monads must eventually admit that the composition of them can also not have any spatial extension and are forced, therefore, to consider both the extension and matter itself as mere phenomena and phantasms, notwithstanding the fact that at the beginning of their reasoning they consider matter to be real; so that, if the conclusion were true, the premises would necessarily have to be false.[204]]

Whether, ultimately, "the mathematician EULER, perhaps alone, must be given the high credit of having, as a result of the monad dispute, through provocation, led philosophy of the mid-18th century decisively out of a blind alley"[205], is probably exaggerated, but EULER's effectiveness, nevertheless, as a "moderate enlightener" — in spite of his strong opposition to certain representatives of the French enlightenment and in spite of his confessional limitation — is beyond doubt; it cannot be ignored in the Western history of ideas.

# Chess

LEONHARD EULER also played chess, and probably not badly. There are some indications of this in the correspondence. In his letter of July 4, 1744 to GOLDBACH, EULER writes from Berlin to Moscow: *Allhier wird stark Schach gespielt: es befindet sich unter andern ein Jud[206] hier, welcher ungemein gut spielt, ich habe einige Zeit bei ihm Lektionen genommen und es jetzt so weit gebracht, daß ich ihm die meisten Partien abgewinne.*[207] [All around here, chess is played passionately: Among others, there is a Jew[206] here who plays extremely well, I took lessions from him for a while and got to the point where I am winning most games with him.[207]] On June 15, 1751 (new style) GOLDBACH, for example, reports to EULER: "I have read in the newspapers already quite some time ago that Mister PHILIDOR[208] has made furore among the greatest chess players in Berlin, from which I surmise that to Your Honorable he is also not unfamiliar"[209], upon which EULER on July 3, 1751 replies by return mail: *Den großen Schachspieler* PHILIDOR *habe ich nicht gesehen, weil er sich mehrenteils in Potsdam aufhielte. Er soll noch ein sehr junger Mensch sein, führte aber eine Maîtresse mit sich, wegen welcher er mit einigen Officiers in Potsdam Verdrüsslichkeiten bekommen, welche ihn genötiget, unvermutet wegzureisen, sonsten würde ich wohl Gelegenheit gefunden haben, mit ihm zu spielen. Er hat aber ein Buch vom Schachspiel in Engelland drucken lassen[210], welches ich habe, und darin gewiß sehr schöne Arten zu spielen enthalten sind. Seine größte Stärke bestehet in Verteidigung und guter Führung seiner Bauern, um dieselben zu Königinnen zu machen, da er dann , wann die Anstalten dazu gemacht, piece für piece wegnimmt, um seine Absicht zu erreichen und dadurch das Spiel zu gewinnen.[211]* [The great chessplayer PHILIDOR I have not seen, because he usually stayed in Potsdam. He is said to be still a young man but kept a mistress, which caused displeasure among some of the officers in Potsdam, who in turn forced him to unexpectedly move away; otherwise, I would probably have had an opportunity to play with him. He had, however, published a book about chess in England[210], which I have, and which certainly contains very beautiful ways of playing chess. His greatest strength is in the defense and in the skillful move of his pawns in order to transform them into queens and, after proper preparation, taking away piece after piece to realize his intention and win the game.[211]]

But let us turn to the mathematical aspects which EULER extracted from the chess board: the problem of the "knight's moves". Here is how our mathematician himself sketched the problem, as formulated in his letter of April 26, 1757 to GOLDBACH: *Die Erinnerung einer mir vormals vorgelegten Aufgabe hat mir neulich zu artigen Untersuchungen Anlaß gegeben, auf welchen sonsten die Analysis keinen Einfluß zu haben scheinen möchte. Die Frage war: man soll mit einem Springer alle 64 Felder auf einem Schachbrett dergestalt durchlaufen, daß derselbe keines mehr als einmal betrete. Zu diesem Ende wurden alle Plätze mit Marquen belegt, welche bei der Berührung des Springers weggenommen wurden. Es wurde noch hinzugesetzt, daß man von einem gegebenen Platz den Anfang machen soll. Diese letztere Bedingung schien mir die Frage höchst schwer zu machen, denn ich hatte bald einige Marschrouten gefunden, bei welchen mir aber der Anfang mußte freigelassen werden. Ich sahe aber, wann die Marschroute in se rediens wäre, also, daß der Springer von dem letzten Platz wieder auf den ersten springen könnte, alsdann auch diese Schwierigkeit wegfallen würde. Nach einigen hierüber angestellten Versuchen habe ich endlich eine sichere Methode gefunden, ohne zu probieren, soviel dergleichen Marschrouten ausfindig zu machen als man will (doch ist die Zahl aller möglichen nicht unendlich); eine solche wird in beigehender Figur vorgestellt:*

| 54 | 49 | 40 | 35 | 56 | 47 | 42 | 33 |
|----|----|----|----|----|----|----|----|
| 39 | 36 | 55 | 48 | 41 | 34 | 59 | 46 |
| 50 | 53 | 38 | 57 | 62 | 45 | 32 | 43 |
| 37 | 12 | 29 | 52 | 31 | 58 | 19 | 60 |
| 28 | 51 | 26 | 63 | 20 | 61 | 44 | 5  |
| 11 | 64 | 13 | 30 | 25 | 6  | 21 | 18 |
| 14 | 27 | 2  | 9  | 16 | 23 | 4  | 7  |
| 1  | 10 | 15 | 24 | 3  | 8  | 17 | 22 |

*Der Springer springt nämlich nach der Ordnung der Zahlen. Weil vom letzten 64 auf Nr. 1 ein Springerzug ist, so ist diese Marschroute in se rediens. Hier*

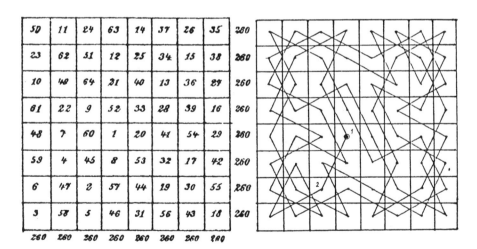

The "magic knight's move" of Jaenish, 1859. Central symmetry becomes evident on account of the line segments of the moves from center to center of the squares – beginning with the square d4.

*ist noch diese Eigenschaft angebracht, daß in areolis oppositis*[212] *die differentia numerorum allenthalben 32 ist.*[213] [The recollection of a problem once posed to me gave rise, recently, to nice investigations upon which analysis usually does not seem to have any bearing. The question was: A knight is to be moved over all 64 squares of a chess board in such a manner that it enters no square more than once. To this end, all squares were covered with tokens which, when touched by the knight, were taken away. It was stipulated, further, that one should start at a given square. This last condition seemed to make the problem exceedingly difficult for me, since I soon found a few marching paths for which, however, the starting point had to be left up to me. I saw that, if the marching path were closed, that is, if the knight were able to move from the last square back to the first, these difficulties would disappear. After a few trials in this regard, I finally found a secure method, without trial and error, to discover as many such marching paths as one wishes (but the number of all possible ones is not infinite); one such path is presented in the figure above.

The knight, in fact, moves as indicated by the numbers. Since from the last, 64, to no. 1 is a knight's move, this marching path is closed. Here, one might still note the property that for opposite squares[212] the numerical difference is always equal to 32.[213]]

So much for EULER. The problem was not new at all — its beginnings extend back to the first half of the 14th century — but it was new for EULER and the respective circle of society. Problems regarding the knight's moves, at that time, were treated even rather frequently in writings and correspondences of important mathematicians.[214] EULER, however, systematizes and generalizes the problem (1759) in a serious memoir[215] and therein solves one of the first problems of combinatorial topology. In view of EULER's later intensive occupation with magic squares[216], one can well imagine how delighted EULER must have been about the "magic knight's move" of the Russian chess theoretician JAENISH[217], in which the position numbers have the constant row- and column sum 260. One has to remark, however, that this "knight's move in holiday clothes" (LANGE) is only semimagic, since the diagonal sum is not also equal to 260. On the other hand, it has the advantage, first, of being "closed" in the sense of EULER, and second, of being symmetric, since the chain 1–32, by a rotation of the board by 180°, is brought into the one from 33 to 64. Up to now, 242 such "magic knight's moves" are known, some of which are open, others closed. The knight's move communicated by EULER to GOLDBACH of course is not magic, but at least symmetric, and EULER was aware of this.

# The great Trio

## FREDERICK II and MAUPERTUIS

Without doubt, even before his accession to the throne in May of 1740, FREDERICK reached a decision to found a new Academy of Sciences and Letters, with Berlin as its center, and in fact he already had looked around in Europe for appropriate and renowned scholars. Since 1736 he carried on a correspondence with VOLTAIRE whom he greatly idolized and to whom he promised the presidency of the academy by way of a hint. HARNACK, however, warns against taking this idea of FREDERICK all too seriously, since FREDERICK knew very well that VOLTAIRE at that time would not have separated from his lady friend, the learned Marquise DU CHÂTELET, and that the latter would not have come to Berlin in any case.[218] So the king considered in earnest the appointment of two very renowned scientists of their

VOLTAIRE

PIERRE-LOUIS MOREAU DE MAUPERTUIS.
Engraving by BONINI after a drawing
by DEMARCHI

time, who as a pair were to set up and direct the academy: The philosopher CHRISTIAN WOLFF (cf. Chap. 2) and the mathematician PIERRE-LOUIS MOREAU DE MAUPERTUIS. FREDERICK already as crown prince became acquainted with WOLFF's philosophy through ULRICH FRIEDRICH VON SUHM, and the connection with MAUPERTUIS probably came about even through his rival VOLTAIRE.[219] Therefore, the young king invited MAUPERTUIS and WOLFF simultaneously; the latter, however (once bitten, twice shy), declined with thanks and requested solely to be appointed as professor and vice chancellor of the University of Halle[220], which was graciously granted to him by FREDERICK. MAUPERTUIS, on the other hand, was inclined to accepting the call. The first tête-à-tête of FREDERICK with the almost designated president – and also the first encounter with VOLTAIRE – came about in September 1740 at the castle Moyland near Kleve, during which VOLTAIRE went to extreme lenghts to discourage MAUPERTUIS from accepting the Academy presidency. "But MAUPERTUIS followed the monarch to

JOHANN II BERNOULLI. Oil painting by
J. R. HUBER (?)

Berlin, whereas VOLTAIRE returned to his marquise. He then already has
played a double game: He simultaneously had hopes of attaining the presidency of the new Berlin Academy and the position of a French ambassador
at the Prussian court. Never has he forgiven MAUPERTUIS for going to Berlin
against his advice and will."[221]

The importance of MAUPERTUIS for the Berlin Academy and, in particular, for LEONHARD EULER justifies a short biographical insertion, even
though much has already been written about him.[222] At the time of his
nomination for president of the Berlin Academy, MAUPERTUIS could already look back on a long academic career in Paris. Born in Saint-Malo
(Bretagne) on September 28, 1698, he enjoyed — more pampered by his
mother than loved—the customary aristocratic education. After an officer's
training in the French army, and studies absolved in Paris — first in music,
then in mathematics and mechanics — MAUPERTUIS was elected in 1723 as
*membre adjoint* into the Paris Académie des Sciences with a dissertation on
the form of the musical instruments, and later (as of 1731) he was on secure pay as *pensionnaire géomètre*. In the following years he devoted himself
predominantly to mathematical problems in the theory of curves, differ-

JOHANN SAMUEL KÖNIG. Oil painting by
GARDELLE

ential geometry, and its application to problems in mechanics, but soon
he turned to geodesy and astronomy, where his rather solid mathematical
education came in useful for him. On the occasion of his trip to England
in the year 1728, MAUPERTUIS became acquainted with NEWTON's theories
and cosmology, and he became one of their most successful promoters
on the continent[223] where DESCARTES's vortex theory was still largely pre-
dominant. Subsequently he travelled to Basel where from September 1729
until mid 1730 he expanded his mathematical knowledge with JOHANN
BERNOULLI, where he also studied NEWTON's *Principia* in 1734 and thereby
counted BERNOULLI's son JOHANN II (JEAN), ALEXIS-CLAUDE CLAIRAUT and
JOHANN SAMUEL KÖNIG among his youthful fellow students. With the first
he kept up a lifelong friendship[229], but KÖNIG was to play a fatal role still
later in MAUPERTUIS's life.

The most important scientific accomplishment of MAUPERTUIS, which
indeed earned him worldwide fame, is without doubt the spectacular success of the 16-month long Lapland expedition of 1736/37. By order of
king LOUIS XV the Paris Academy in 1735 sent a research expedition to
the equator into what is today Ecuador in order to settle the dispute, by
means of a geodesic-astronomical measurement of latitude and longitude,

as to whether our planet represents an ellipsoid elongated toward the poles, as the French astronomer CASSINI claimed, or rather a flattened one, as had to be concluded from NEWTON's theory. The original intention was, quite openly, to validate CASSINI and contradict the Englishman. This expedition took place under the auspices of its initiator LOUIS GODIN[225] who was accompanied by BOUGUER, DE JUSSIEU, and LA CONDAMINE[226]. It lasted ten years and did not run a particularly happy course; besides, it did not produce the result that was actually hoped for.

This is where MAUPERTUIS came in. He proposed to the Academy that the dispute could be decided more quickly, more simply, and cheaper through an appropriate measurement near the north pole. Flattening at the poles namely would have the consequence of the arc length measured along a meridian to an astronomically determinable latitude turning out to be larger near the pole than, say, in central Europe, or even at the equator. Therefore, the Paris Academy in 1736 started an additional expedition, namely to Lapland, under the direction of MAUPERTUIS, accompanied by CLAIRAUT, LE MONNIER, OUTHIER, CAMUS, and the Swedish physicist CELSIUS with whom everyone today is familiar through the temperature scale named after him. The trigonometric measurements of the exceedingly wearing Lapland expedition indeed yielded a clear enlargement of the meridian arc length of a latitude, and thus the first empirical proof for the flattening of the earth at the poles. This flattening could then be determined quantitatively in 1745 through comparison with measurement results in Peru.

MAUPERTUIS was triumphant and let himself be celebrated in Paris as *aplatisseur de la terre* [flattener of the earth]. "... in this way the victory was decided, France definitely broke away from DESCARTES and acknowledged unanimously: NEWTON is great, and MAUPERTUIS is his prophet! Those were the best days in the life of MAUPERTUIS. His name was on everyone's tongue, he himself was courted and idolized in the salons of the high society as the hero of the hour. Indeed, owing to him, mathematics became fashionable in the Parisian society ... His travel report[227] was devoured, he himself celebrated as a second NEWTON, as a new JASON. That's the way France's greatest poet VOLTAIRE sang his praises, in the same breath comparing him with ARCHIMEDES, COLUMBUS, and MICHELANGELO! In the same way he was viewed by FREDERICK, who liked the witty and self-confident man."[228] He

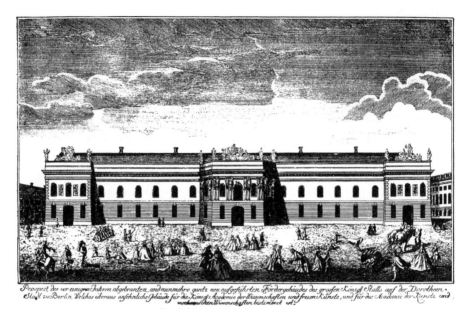

*Prospect der vor einigen Jahren abgebranten, und nunmehro gantz neu aufgeführten Vordergebäudes des großen Königl Stalls auf der Dorotheen-Stadt zu Berlin. Welches überaus ansehnliche Gebäude für die Königl Academie der Wissenschaften und freuen Künste, und für die Academie der Künste und mechanischen Wissenschaften bestimmet ist.*

123. **Die Akademie der Wissenschaften.** 1752
Nach einem Stiche von Schleuen

The building of the Berlin Academy, 1752

liked him so much that he had him come to his camp in Reichenbach in March of 1741, after he had celebrated together with him and VOLTAIRE in November 1740 the famous "joyous festivals in Rheinsberg", and had marched into battle in December, although during the already raging First Silesian War this was certainly not the time to think about the formation of the new academy. Yet MAUPERTUIS (as an old soldier) accepted the invitation obviously with pleasure, and the rather odd incidents that occurred after his capture in the battle at Mollwitz on April 10 — as is known declared lost prematurely by FREDERICK II — retain to this day an undeniably anectodal appeal: MAUPERTUIS was not immediately recognized, robbed, but ultimately transferred to Vienna and presented to the Empress MARIA THERESA with all honors.[230] From Vienna, MAUPERTUIS went to Berlin for a short time, before EULER was present, and then on to Paris to do scientific work for France — without, however, taking away the Prussian king's hope for his return to Berlin. In the year 1742, MAUPERTUIS becomes the director of the Académie des Sciences, and one year later — on the recommendation of MONTESQUIEU — is even elected to be one of the "Forty Im-mortals". On June 27, 1743 he gave his inaugural speech in the Académie Française, in which he compared the activities of the mathematician to those of the poet and orator. — MAUPERTUIS was at the zenith of his fame. Meanwhile, in mid-1742 FREDERICK II had brought the First Silesian War to an end acceptable to him with the peace treaty of Breslau, but still — de-spite EULER's urging — the king did not give any thought to officially unite the Old Society with the new Société Littéraire to form the new Academy, even though the financial basis to do so was to some degree assured by the massive increase in calendar sales as a main source of income for the academic societies in the conquest campaign of Silesia: He did not want to found the Academy without "his MAUPERTUIS", and the latter, for his part, was waiting for more peaceful and more secure times in the kingdom of Prussia. But these times were not to be yet, for in the spring of 1744 FREDERICK immersed himself in the Second Silesian War, for reasons of supposed security and the expansion of the conquered regions, and until after the battle at Hohenfriedberg in June of 1745, victorious for Prussia, nothing more is heard about his academy project, since the end of the war was long in coming with the peace treaty of Dresden on Christmas day.[231] During this half-year, FREDERICK II, after in June he had received from MAUPERTUIS, to his great joy, the definitive confirmation that he now was permitted to come to Berlin, wrote no fewer than sixteen letters with the intent of keeping a hold on him.[232] This, however, would have been un-necessary, because MAUPERTUIS proceeded to Berlin already in the summer to marry ELEANOR VON BORCKE, a daughter of the Major General FRIEDRICH WILHELM VON BORCKE[233]. Early in January 1746, FREDERICK finally returns to Potsdam, gives MAUPERTUIS firm employment with 3,000 Taler yearly salary, and vests the new academy president with all conceivable kinds of authority, among which the stipulation that MAUPERTUIS is to be the superior of all members, including the honorary members, and he is to award pensions (i.e., salaries) solely at his own discretion. "In this form did the statutes appear on May 10, 1746 and were read at the session of June 2; they subjected the Academy to the nearly autocratic power of the new president."[234] Already in the second session of the Academy (on June 9, 1746), VOLTAIRE and LA CONDAMINE were proposed by MAUPERTUIS as foreign members, and also immediately and unanimously elected, and two

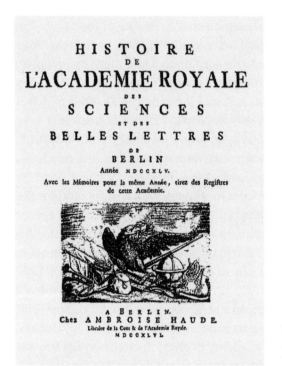

HISTOIRE
DE
L'ACADEMIE ROYALE
DES
SCIENCES
ET DES
BELLES LETTRES
DE
BERLIN
Année MDCCXLV.
Avec les Mémoires pour la même Année, tirez des Regiftres
de cette Académie.

A BERLIN,
Chez AMBROISE HAUDE.
Libraire de la Cour & de l'Académie Royale.
MDCCXLVI.

Frontispiece of the first volume of the Berlin Academy journal for 1745, published in 1746

weeks later "MAUPERTUIS announces that FREDERICK II has declared him-self the protector of the Academy. The king consulted with the president on all important matters regarding the Academy, and also in the Academy, as goes to show from his self-appointment as protector, he appeared as an enlightened despot."[235]

In exact accordance with this image is the fact that FREDERICK already in July 1745 decreed to the Academy that the publications were all to appear in the French language. Should the author wish, the original could be printed alongside in a "foreign language" (this "gracious permission" has never been made use of, however).

## MAUPERTUIS and EULER

MAUPERTUIS may well have heard a song of praise about EULER first in Basel from the BERNOULLIS, then from CLAIRAUT and KÖNIG. He opens the correspondence[236] with his letter of May 20, 1738, in which he expresses his admiration for EULER's *Mechanica*, thanking him for the work, and reciprocating with his *Figure de la terre*. Because of his polar expedition, MAUPERTUIS didn't get around to reading EULER's epoch-making principal work of 1736 until the winter of 1737/38. The two met personally for the first time in Berlin in January 1745, albeit under circumstances that did not seem to allow any serious contact yet: The continuous festivities at the court, from which EULER stayed away, kept MAUPERTUIS from "meeting with him as often as I would have wished"[237]. This was to change only with the start of the regular Thursday sessions of the Academy, the first of which took place on June 2, 1745.[238]

As a scientist, MAUPERTUIS naturally could by far not hold a candle to EULER, and both scientists — in tacit agreement — were very aware of this. In their character and personality they were quite different[239]: MAUPERTUIS arrogant, extremely vain, excessively hurt and hurtful, at times inconsiderate, whenever it came to asserting the prestige of his position and defending supposed priorities, as the case of König (1751) with its ugly quarrel about the "principle of least action" should prove particularly impressively[240], in the course of which, though, also EULER had to take, by VOLTAIRE's sharp pen, several rather unsightly spatterings onto his otherwise so clean moral slate. We mean here his completely incomprehensible — that is, to this day not adequately explainable — attitude toward his pleasant compatriot KÖNIG in the dispute about the Leibniz–Euler resp. Euler–Maupertuis "principle", where he, together with MAUPERTUIS, supported only by FREDERICK II — and disavowed by his Swiss colleagues in the Academy — believed to have to stand alone, isolated as it were, against the entire academic republic, led by VOLTAIRE, and condoned the "academic judicial murder", in fact even fostered it. This probably ugliest and most spectacular quarrel in the history of science of the 18th century had, to be sure, only a negligible effect on EULER's life, yet the poison splattered by VOLTAIRE in his *Diatribe du Docteur Akakia, médecin du Pape* (1753) was to nearly send the academy president MAUPERTUIS to an early grave. MAUPER-

TUIS died in 1759 at the home of his friend JOHANN II BERNOULLI in Basel and was buried in the nearby (catholic) village of Dornach.

Consensus between them existed, without doubt, with regard to the appropriateness of philosophical and metaphysical questions in the program of the Academy, as well as in certain fundamental questions regarding faith and religion in opposition to the radical "free-thinkers" in the vein of French Enlightenment. The relationship between EULER and MAUPERTUIS has been aptly sketched by Winter: "EULER clearly is the leading force in this relationship, and he tries to assert the influence he has on MAUPERTUIS also with full success for the development of the Academy. This secret presidency allows EULER to get over the fact that a scholar beneath his scientific level ... stood above him at the helm of the Academy. How difficult it was for EULER, however, to put up with this subordination becomes evident from a letter he had written to his old friend GOLDBACH in Petersburg already one year prior. From this letter there resonates his true attitude toward the French, who exerted their influence in Berlin, about the high esteem [,] in which they were held, which however was in no way commensurate with the intellectual weight that was due them."[241]

How MAUPERTUIS for his part felt about EULER's character can be gleaned from one of his intimate letters to JOHANN II BERNOULLI: "EULER ... is all in all an exceedingly peculiar personality, a relentless pest, who likes to meddle in all affairs, even though the form of our Academy and the directives of the king do not permit anyone any kind of meddling ... This is yet another of those family secrets that one should not pass on."[242] This remark is certainly not pure invention, yet it sheds a still brighter light on MAUPERTUIS himself, who benefitted to a large measure by just this quality of EULER, as the *Registres* and the correspondence unmistakably show.

## EULER and FREDERICK II[243]

The letter of the Prussian king to EULER, mentioned on p. 60, was not the first one regarding the matter of his appointment that had been delivered to him by the ambassador VON MARDEFELD, for we read in EULER's letter to MÜLLER dated February 23/March 6, 1741: *Ungeacht auch anjetzo der König mit der Conquete von Schlesien beschäftiget, so hat er dennoch die höchste Gnad*

Leonhard Euler. Oil painting of E. Handmann, 1756

*gehabt, schon etliche Mal eigenhändig meinetwegen hieher zu schreiben: und ich erwarte jetz alle Tage die letsten Ordres Ihro Majestät und meine Demission, um meine Abreise vorzunehmen.*[244] [Regardless of the fact that the king also at this time is occupied with the conquest of Silesia, he still has with utmost graciousness written to me several times, and in his own hand: And I am expecting any day now the latest instructions from His Majesty and my dismissal in order to get on with my departure.[244]] With the letter of King FREDERICK out of the "Camp of Reichenbach" of September 4, 1741 begins the correspondence[245] between these two so prominent personalities, that was to last a quarter century. Both have made world history: One in the domain of science, the other in the area of politics. As different as their field of activity was, as disparate were also their character and their education. On the one hand, a truly brilliant thinker and mathematician who has decisively driven forward the development of his science, the modest son of a country minister, unassuming and inconspicuous in appearance; on the other hand, the glittering figure of a young king, veiled in glory of war, who loved nothing more than sparkling and witty salon conversation — contrasts, as one could not imagine them to be more grotesque. A personal meeting between FREDERICK II and EULER took place for the first time on September 6, 1749 at the court in Potsdam on the occasion of a hydro-technical assignment (water games in Potsdam) that EULER had been given by the king.

FREDERICK's assessment of LEONHARD EULER in particular, and of science in general, becomes apparent, for example, through the correspondence with his brother AUGUST WILHELM. The prince wrote to the monarch on October 28, 1746: "Mr. MAUPERTUIS has introduced me to the mathematician EULER. I found in him the truth confirmed of the imperfection of all things. Through industriousness he has acquired logical thinking and thereby a name for himself: But his appearance and awkward expression obscure all these beautiful characteristics, and they prevent us from benefitting from them", and FREDERICK replied on October 31 of the same year: "Dearest Brother! I was afraid that your talk with Mr. EULER would not uplift you. His epigrams consist of calculations of new curves, some sorts of conic sections, or of astronomical measurements. Among the scholars there exist such monumental calculators, commentators, translators, and compilators[246], who are useful in the republic of the sciences, but otherwise are anything but brilliant. They are used as are the Dorian columns in architecture. They belong to the subfloor, as support for the whole edifice and the Corinthian columns which form its adornment."[247]

The princely gentlemen could not have given themselves a more succinct testimonial on their intellectual insufficiency, and indeed in FREDERICK's complete lack of appreciation for all things mathematical there lies one of the more profound reasons for his inexcusable misbehavior toward EULER, which later was to result in the latter's return to Petersburg. That this attitude represented not only the occasional expression of a whim in wanting to appear clever, but rather a deep-seated disregard, already present in the crown prince, for all that he did not comprehend, is attested to in his letter to VOLTAIRE of July 6, 1737, in which one can read something to the effect that a king was to run an academy as would a squire keep a pack of dogs. If we refer to FREDERICK's misbehavior toward EULER with the harsh label "inexcusable", we must perhaps amplify it with "yet understandable", since "FREDERICK II in December 1745 was in the prime of his life but was already burdened by many ailments. He suffered from gout and colics. Hemorrhoids were troubling him. His digestive system was not intact. Bouts of fever forced him frequently to bedrest. The Prussian king at one time later declared the gout as 'inherited from his father'. In reality, he had inherited far more from him, namely that horrible metabolic disorder that at times had driven FREDERICK WILHELM I to the edge of insanity ... FREDERICK, at that time, expected only a short life. Twelve years is all he gave himself ... He felt indeed worn out. His view of his own chances for survival became reinforced by a minor stroke he suffered in the year 1747 – at the age of thirty five."[248] Whether EULER was aware of this can be assumed, however not verified. What can be verified easily[249], though, are the manifold, practical domains in which the mathematician was active "on the subfloor" for "his king" during more than twenty years. A concise listing of key words must suffice here: Levelling of the 70 kilometer long Finow canal with seventeen locks, linking the Oder with the Havel, of the greatest importance for the connection on water between Berlin and Stettin (EULER took his fifteen-year-old son JOHANN ALBRECHT along on this trip); dam and bridge constructions in the principality of Eastern Friesland, fallen to Prussia in 1744; grading of the brine in the salt mine of Schönebeck; measures for the drainage of the Oder marshland;

calculations and installation supervision of the water games at Sanssouci (pumps and hose technology); calculation and construction of horse and wind mills, as well as of water scoop equipment; supervision of wall repair work in the Botanical Garden and installation of the green houses of the mulberry plantations for increasing silk production; improvement in minting medallions and coins; rationalization of calendar sales, as well as assessment of Italian and Dutch lotteries (something that actually rather belonged again to the job of the mathematician).

Yet, concurrent with all these activities, EULER conducted his purely scientific research. At the time of his Finow travel (1749), there appeared his epochal study *On the perfection of the telescope objectives*[250]. This gives us cause to insert a section on the significance of optics in EULER's work.

## Optics

The fundamental work was announced by EULER in a letter to GOLDBACH of October 1748 with the following words, outlining the problem very clearly:... *und anjetzo bin ich bemühet, einen Einfall ins Werk zu richten, welchen ich gehabt, um solche Objektiv-Gläser zu verfertigen, welche eben den Dienst leisten sollen, als die Spiegel in den tubis Neutonianis und Gregorianis*[251]. *Der Fehler der gewöhnlichen Objektiv-Gläser rühret nur daher, daß die Licht-strahlen nicht einerlei Refraktion leiden und also zum Exempel die roten Strahlen einen andern focum formieren als die blauen, dahingegen von einem Spiegel alle Strahlen in eben denselben focum reflektiert werden. Dieser Unterschied zwischen den focis der roten und blauen Strahlen wird auch um so viel größer, je weiter dieselben vom Glas entfernet sind, und bei einem Objektivglas von 27 Schuh fällt der rote focus einen ganzen Schuh weiter als der blaue, woraus die Undeut-lichkeit und die Farben der durch lange tubus gesehenen objectorum entspringen. Wann man also solche Objektivgläser verfertigen könnte, welche alle Strahlen in einen gemeinsamen focum zusammen würfen, so würde man von denselben eben diejenigen Vorteile zu gewarten haben, als von den Spiegeln. Dieses ist aber nicht möglich mit blossem Glas zu bewerkstelligen. Dahero bin ich auf den Gedanken gefallen, ob es nicht möglich wäre, aus Glas und Wasser oder zwei anderen ver-schiedenen durchsichtigen Materien solche lentes objectivas zu verfertigen, und*

*zweifelte hieran um so viel weniger, da wir sehen, daß in den Augen, welche
aus verschiedenen durchsichtigen Körpern bestehen, eine solche Undeutlichkeit
wegen der verschiedenen Brechung der Lichtstrahlen nicht wahrgenommen wird.*
[... and currently I am trying to work out an idea I had for building such
object glasses which should do the same service as do the mirrors in the
reflector telescopes of Newton and Gregory. The error in the usual object
glasses stems only from the fact that the light rays do not undergo the
same refraction and thus, for example, the red rays form a different focus
than the blue ones, whereas with a mirror all rays are reflected into one
and the same focus. This difference between the foci of the red and blue
rays also becomes larger the farther apart the latter are from the glass, and
for an object glass of 27 feet, the red focus comes to fall a whole foot farther than the blue one, which is the cause for the blurriness and the colors
of objects seen through a long telescope. If one thus could build object
glasses which would reflect all rays into a common focus, one would ex-
pect from them the same advantages possessed by mirrors. This, however,
is not possible to accomplish with glass alone. Therefore, the idea came
to me whether it should not be possible to build such object glasses using
glass and water, or two other different transparent materials, and had all
the less doubt about it as we see that in the eyes, which consist of different transparent substances, such blurriness owing to different refraction
of light rays cannot be discerned.] {There follows a sketch and description
of a glass-water-triplet}.[252]

The discourse here thus is about the elimination of the chromatic error
of lenses, about "achromatism", at that time a new research area of geometric and physical optics. Questions of optics in the widest sense have
occupied EULER throughout his life, and nowhere more clearly than in
this science is the contrast with the school of NEWTON more evident. It is
not easy to decide to what extent this fact may have influenced GOETHE's
exceedingly favorable (and often cited) judgment: "EULER, one of those
men who are predestined to again start from scratch even if they find
themselves in a harvest, however rich, of their predecessors, did not put
out of his mind the consideration of the human eye, which by itself does
not see apparent colors, even though it sees and becomes aware of objects
through significant refraction, and he had the idea of joining together
meniscal lenses, filled with different moistures, and through experiments

and calculations came to the point where he dared to assert that the color phenomenon can be eliminated in such cases, and there would still remain refraction."[253]

Already in one of his first writings on optics[254], EULER confronted NEWTON's corpuscular theory, whose main protagonists in Paris in the first generation after EULER were to be LAPLACE and BIOT, with a wave theory of HUYGENS's stamp, but opposition against the emission theory in England itself was slow in coming: With the exception of ROBERT HOOKE, the older contemporary of NEWTON, the first English physicist of note who openly spoke out against the emission theory was THOMAS YOUNG in his "Bakerian Lecture" — though with the weighty argument which was not yet at EULER's disposal: a theory of interference[255]. As a matter of fact, interference, refraction, and polarization phenomena cannot be explained in an entirely satisfactory manner, the two latter not at all, by a wave theory oriented solely longitudinally, and EULER, strangely enough — in spite of his intense occupation with the vibrating string — did not hit on the idea of the transversal vibration of light.

Ever since the use of refracting telescopes by GALILEI and HARRIOT at the beginning of the 17th century, the rings of color in the image field, inevitable in those times, proved to be quite disturbing, which is the reason why also GREGORY and NEWTON switched to the reflector telescope, more favorable in this respect. This error in color, the chromatic aberration, is a direct consequence of the fact that light of different wave lengths, that is, of different colors, is unequally refracted in the same medium. Only on the basis of NEWTON's investigations of the dispersion of light in a prism could one begin to consider the possibility of eliminating the chromatic error. NEWTON himself originally thought it impossible to achieve achromatism by means of variably refracting media because of insufficient experimental data, and initially also the optician JOHN DOLLOND in London, well known in the history of technology, until he succeeded, after a few wrong tracks, to build an achromatic device using a combination of crown glass and flint glass lenses.

EULER's contribution to this discovery is considerable, since DOLLOND had been influenced decisively by EULER's work, which appeared in 1749, as well as by a memoir of the Swede SAMUEL KLINGENSTIERNA, which in turn had been stimulated by the work of EULER. Most remarkable, how-

# DIOPTRICAE
### PARS PRIMA
CONTINENS
### LIBRVM PRIMVM,
DE
### EXPLICATIONE
# PRINCIPIORVM,
EX QVIBVS
### CONSTRVCTIO TAM TELESCOPIORVM
QVAM
### MICROSCOPIORVM
EST PETENDA.

---

AVCTORE
### LEONHARDO EVLERO
ACAD. SCIENT. BORVSSIAE DIRECTORE VICENNALI ET SOCIO
ACAD. PETROP. PARISIN. ET LOND.

*PETROPOLI*
Impenſis Academiae Imperialis Scientiarum
1 7 6 9.

Frontispiece of EULER's universal "Optics", Petersburg 1769

ever, is EULER's (wrong) main argument, on which he based his belief in the possibility of achromatism: the supposed absence of chromatic error in the human eye. Even though NEWTON already pointed out the chromatic error of the eye, which can be detected by partially dimming the pupil[257], EULER not only held on to the idea of the absence of chromatic error in the eye, but even wanted to see therein a sure indication for the existence of God[258]. The statement, often made, that EULER invented the achromatic telescope, is incorrect. The first achromatic device (crown and flint glass) one owes to CHESTER MOOR HALL, who seems to have made his (accidental?) discovery around 1729, and DOLLOND successfully reinvented it in 1758. It is an established fact that EULER at that time calculated and experimented only with the optical media glass (ordinary crown) and water, but not with crown and flint glass. Nevertheless, one has to credit him with the merit of having, through his publications, brought about the repetition of NEWTON's prism experiment by DOLLOND and thereby having broken the curse of Sir Isaac's authority.[259]

What was said with regard to the method in EULER's *Mechanics* applies in no lesser degree to his three-volume mighty work *Dioptrics*[260] (1769–1771), which — for the larger part still written in Berlin — became for a long time the textbook of geometric optics and was EULER's own synopsis. Contrary to his predecessors, all of whom proceeded synthetically, EULER treated optics analytically by means of the differential and integral calculus. It is true that in his reflection theory of lens systems he limited himself consistently to points on the optical axis, but for these, the opening and color magnification errors he treated so thoroughly as no one else, and in this way, at least the theory of the astronomical telescope was brought to a preliminary conclusion. EULER, however, became victim of a fundamental and fatal error with his assumption that the aberration effect in the case of light incidence oblique to the axis (aplanatic and coma error) is negligible compared to the opening error (spherical aberration). This, in fact, is by no means the case, since all these errors are of the same order of magnitude. This was clearly recognized also by CLAIRAUT and D'ALEMBERT, which, in this research area, provided them with a significant lead over EULER, which the mathematician, meanwhile grown almost blind, did not attempt to rectify. Nevertheless, EULER's insights are astonishing, even if one considers only — in comparison with the famous optics of GAUSS — his *General theory of dioptrics*[261] produced in 1765.

One must also not forget that in EULER's works very often one finds remarks aside from the main topic, which not rarely have stimulated, or even anticipated, later works of other researchers. One could think, for example, of the distinction between *light intensity* and *illumination intensity*, as, somewhat later, can be found in JOHANN HEINRICH LAMBERT's "photometry"; and WILLIAM HERSCHEL, with his calculation of duplets and telescopoculars, in a certain sense continues EULER's dioptrical efforts.

Astonishing, also in the area of optics, is EULER's depth and abundance — the *Optica* takes up seven quart volumes in the collected works — and almost incredible is the fact that a not insignificant part of EULER's late works about light was the work of a blind man.

# Pater familias

Is it possible that LEONHARD EULER was satisfied with his time in Berlin? The fact that he persevered in the Prussian metropolis for a quarter century would seem to suggest to us a clear yes, yet a conclusive, all embracing answer to this question, in view of the complexity of human existence, is probably not possible. Let us illuminate — more or less chronologically — some aspects of EULER's life during the Berlin period before the definitive break with FREDERICK II.

Ever since his departure from Petersburg, EULER drew from the academy there, with which he continued to be in close collaboration, a yearly pension of 200 Rbl. Owing to a decree of the Empress ELISABETH, this payment was terminated for EULER and all the other foreign members in mid-1743 by Nartov[262], who had to substitute as chief chancellor for SCHUMACHER, arrested in the meantime[263]. Although EULER officially, even without pay, was content with his honorary membership[264], he didn't refrain from airing his displeasure in a letter to GOLDBACH: *Der H. Nartow hat mir die Pension, welche mir von der Akademie akkordiert und von der Kommission aufs nachdrücklichste konfirmiert worden, aufgekündet; ein gleiches Schicksal hat alle ausländischen Pensionnaires getroffen, worüber der H. [Daniel] Bernoulli inkonsolabel ist. Weil ich nun von meiner Obligation gegen die Akademie freigesprochen bin, so lasse meine Scientiam navalem[265] bei Mr. Bousquet in Lausanne drucken.*[266] [Mr. Nartov has revoked the pension that had been accorded to me by the academy and strongly confirmed by the commission; the same fate befell all foreign pensioners, about which Mr. [Daniel] Bernoulli is inconsolable. Since I have now been released from any obligation to the academy, I will have my *Scientiam navalem*[265] printed by Mr. Bousquet in Lausanne.][266]

It is probable that EULER, in his first disappointment about the long delayed establishment of the Berlin Academy, owing to the Silesian war, entertained the idea of purchasing a country estate in Switzerland, and to settle in his homeland. This we know from a letter of GOLDBACH from the early summer of 1744: "I remember that Your Honorable have told me already several years ago, in what way you had in mind spending a capital of 10,000 Rth., should you acquire it, namely to purchase a country estate in your native land, and to live on it; but even though termination presumably will soon be the case, I hope, though, that you will not bring to fruition your project from that time."[267]

Although the second Silesian war was still raging, EULER seemed to have given up the "project from that time", as he tells his friend GOLDBACH by return mail: *Meine Umstände haben sich jetzt dergestalt geändert, daß ich töricht sein müßte, wann ich mir eine andere Lebensart wünschen sollte. Wann aber auch dieses nicht wäre, so fehlte mir doch noch so viel an derjenigen Summ von 10 000 Rth., welche ich vormals zur Erkaufung eines Landguts in Patria anwenden wollte, daß ich kaum Hoffnung haben kann, zu derselben jemals zu gelangen.*[268] [My circumstances have now changed in such a way that I would be foolish to wish a different life style for myself. But even if this were not the case, I would still be lacking so much toward the sum of 10,000 Rth., which I previously wanted to use to buy a country estate in the homeland, that I hardly can have any hope to ever attain such.][268]

Immediately after RASUMOVSKI's appointment as president of the Petersburg Academy (1746), EULER, who in February of the same year had been elected foreign member of the London Royal Society, was sent the invitation to return to Petersburg. However, he declined, for at last the Berlin Academy was functioning under the presidency of MAUPERTUIS with EULER as director of the Mathematical Class. All the more surprising is the letter that EULER on March 5, 1748 sent off to London to his friend JOHANN KASPAR WETTSTEIN, chaplain and librarian to the Prince of Wales, and in which one reads, among other things (in free translation): *The newspapers tell us much about the intentions of the* [British] *parliament to naturalize foreign protestants. This is a matter to which I am not indifferent, for I have a large family, and there is no other country in which I would rather like to settle than England. You have shown me so many signs of your goodwill and your friendship, that I cannot help but to fully confide in you regarding this matter, and to ask you for your assistance in this very delicate affair. I am namely observing that here the interest in literature more and more displaces that in the mathematical sciences, so that I have reason to fear that soon I shall become useless. In such a case I would not want to return to Petersburg, because there, I cannot hope for a secure livelihood for my family. But since the latter is now so numerous, I don't see for myself, neither in our homeland nor elsewhere, a suitable position — except in England.*[269] EULER then adds the request that WETTSTEIN should try to find such an appropriate, exceptional position that would pay him,

EULER, as much as the one in Berlin, and he obliges WETTSTEIN to absolute confidentiality, since no one was supposed to know anything about these intentions. — Is it possible that WETTSTEIN knew that early in 1748 EULER had been offered the succession of JOHANN I BERNOULLI's chair in Basel? Be it as it may: EULER's rejection — quite apart from the reasons mentioned earlier in the Prologue — is readily understandable, considering the salary rates then customary in Basel.[270]

With regard to important facts from the private sphere during those Berlin years, a few can be noted, as, for example, KATHARINA EULER's twin birth of two girls in April 1749, who, however, were to die already in August of the same year. Also LEONHARD EULER's twenty-day trip to Frankfurt a. M. and back, when in June 1750 he went to get his mother, widowed since 1745, to live with him in Berlin. Furthermore — as a curiosity — his lottery winnings of 600 Rth., *welches ebensogut ist, als wann ich dieses Jahr* [1749] *einen Pariserpreis gewonnen hätte*[271] [which is just as good as if I had won a Paris prize this year][271], and finally, he bought in 1753 *ein sehr angenehmes Landgut in Charlottenburg für 6000 Rth. ... , wobey sehr viel Korn und Wiesewachs befindlich. Daselbst wird meine Mutter, Schwiegerin und alle Kinder nebst einem Hofmeister wohnen, wodurch ich meine bissher sehr starke Haushaltung mehr als um die Hälfte erleichtern und von dem Gut alles, was wir hier brauchen, ziehen kann.*[272] [a very comfortable country estate in Charlottenburg for 6,000 Rth. ... , on which property there is a great amount of corn and meadowland. That's where my mother, sister-in-law, and all the children along with a steward will live, through which I will be able to lighten my household, very large up until now, by more than half and draw from the estate everthing that we need here.][272] This estate, by the way, was mistakenly looted in 1760 during the occupation of Berlin by Russian troops in the Seven-Year War: *Daselbst ist mir nun ... gar alles zerstöret und verheeret worden, indem ich die Salvegarde, welche mir Sr. Erlaucht, der H. Graf TSCHERNISCHEFF, darüber zu ertheilen geruhet haben, zu späth erhalten, dann sobald sich unsere Armée zurückgezogen hatte, wurde Charlottenburg sogleich den Cosaken preissgegeben. Ich habe dabey 4 Pferde, 12 Kühe, eine Menge Kleinvieh, viel Haber und Heu eingebüßet, und in dem Hause sind alle Meubles ruinirt. Dieser Schaden, aufs genauste berechnet, beläuft sich über 1100 Rth., wofür ich 700 Roub.[!] ansetzen will, nicht zu gedenken, daß ich völlig außer Stand gesetzt worden, mein Feld auf das künftige Jahr bestellen zu lassen, daß ich meinen gantzen Schaden nicht unter 1200 Rub. rechnen kann.*[273] [There, almost everything has been destroyed and devastated, because I have received the security guard that His Lordship, Count CHERNISHEV has graciously granted me, too late, for as soon as our army had retreated, Charlottenburg was immediately surrendered to the Cossacks. I thereby have lost 4 horses, 12 cows, a good many small animals, a lot of oats and hay, and in the house all the furniture is ruined. This damage, on most accurate calculation, amounts to more than 1,100 Rth., which I will place at 700 Rbl., not to speak of the fact that I was completely deprived of having my field cultivated for the coming year, so that I cannot estimate my entire damage below 1,200 Rbl.][273]

EULER was more fortunate with his son JOHANN ALBRECHT.[274] The latter, only twenty years old, at the request of MAUPERTUIS became in December 1754 a member of the Berlin Academy — admittedly still in dependency of his father: *Für den Hochgeneigten Anteil, welchen Ew. Hochwohlgeb. an der Beförderung meines ältesten Sohns, als Dero gehorsamsten Patens zu nehmen belieben, erkenne ich mich auf das schuldigste verpflichtet und erkühne mich, Denselben noch ferner Dero Huld und Gewogenheit gehorsamst zu empfehlen. Er wendet allen möglichen Fleiß an, sich der besonderen Gnade, welche ihm durch die Aufnahme in unserer Akademie wiederfahren, je länger je mehr würdig zu machen. Es ist aber jetz[t] das mathematische Studium so weitläufig, daß es eine lange Zeit erfordert, ehe man sich in allen Teilen so fest setzen kann, dass man ohne Anstoß etwas namhaftes darin zu leisten imstand kommt; dahero er freilich ohne meine Hülfe noch nichts Sonderliches würde zum Vorschein bringen können. Insonderheit muß er sich ja in keine gelehrte Streitigkeiten mischen, weilen sonsten seine Antagonisten, welche ihm more Hermanniano antworteten, nicht so sehr unrecht haben dürften. Ich habe aber die gute Hoffnung, daß er je länger je mehr Stärke erreichen und meines beständigen Beistandes nicht mehr lang benötiget sein werde.*[275] [For the very kind interest Your Honorable has taken in the promotion of my eldest son as your most obedient godchild, I am most humbly obliged, and I am so bold as to respectfully request to entrust him also in the future to your graciousness and favor. He is applying himself with the utmost diligence to become ever more worthy of the special favor which is bestowed on him through the admission to our academy. But the mathematical field of study today has become so multifaceted that it takes a great deal of time before one is able

Johann Albrecht Euler. Oil painting by E. Handmann, 1756

to gain solid ground in all parts in order to accomplish, without incentive, something substantial therein; for this reason he naturally would not be capable yet to bring forth anything outstanding without my help. In particular, he must not, after all, get involved in any scholarly disputes, since otherwise his antagonists, who would answer him in the manner of HERMANN, might not be so wrong. But I do have the good hope that over time he will become stronger and will not need my constant assistance much longer.][275] The penultimate sentence refers to a rather subtle postscript of the preceding letter of GOLDBACH to EULER: "The promotion of your eldest son, on which I congratulate Your Honorable as well as him from my heart, I have recently learned about with much joy. I am certain that he has already acquired extraordinary knowledge in mathematics, nevertheless he will still have to endure, in case he is a lover of scholarly disputes, to be contradicted by his antagonists in the manner of HERMANN as LEONHARD EULER rather than as LEONHARD EULER's son."[276]

In the spring of 1760, JOHANN ALBRECHT EULER married ANNA CHARLOTTE SOPHIE HAGEMEISTER of the same age. Only at the end of the Seven-Year War (1763) — the marriage had already brought forth two children — did the increase from 200 to 400 Rth. in yearly salary, promised to JOHANN ALBRECHT EULER since 1760, take effect, alleviating somewhat the son's material dependency on the father. Contributing to this, without doubt, was the fact that the young EULER in 1762 was awarded — jointly with CLAIRAUT — a prize of the Petersburg Academy for his (?) work on the "Störungen der Kometenbewegung durch die Planetenattraktion" [Perturbations in the motion of comets by the attraction of the planets].[277]

It appears that LEONHARD EULER was a very caring father. In the spring of 1761 he accompanied his son KARL to Halle, where the latter was to undertake the study of medicine and receive his doctorate in the fall of 1763. On this occasion, EULER stayed at the house of the very important scholar JOHANN ANDREAS SEGNER[278], with whom for 20 years he had been in active correspondence that was to continue until the death of the latter.[279] The third son CHRISTOPH "widmete sich dem Kriegswesen" [dedicated himself to warfare], and at the age of nineteen was a lieutenant in the artillery. Daughter CHARLOTTE became engaged in 1763 to the young, very wealthy Baron JAKOB VAN DELEN from Holland, at the time an officer cadet in the

Leonhard Euler's children

(The three underlined sons outlived their father.)
Ptb.=Petersburg

| | | | | |
|---|---|---|---|---|
| 1. Johann Albrecht | *27.11.1734 | Ptb. | † 18.9.1800 | Ptb. |
| 2. Anna Margaretha | * 8.6.1736 | Ptb. | † 2.7.1736 | Ptb. |
| 3. Maria Gertrud | * 9.5.1737 | Ptb. | † 1.5.1739 | Ptb. |
| 4. Anna Elisabeth | * 5.11.1739 | Ptb. | †19.11.1739 | Ptb. |
| 5. Karl Johann | * 15.7.1740 | Ptb. | † 16.3.1790 | Ptb. |
| 6. Katharina Helene | *15.11.1741 | Berlin | † 4.5.1781 | Wiborg |
| 7. Christoph | * 1.5.1743 | Berlin | † 3.3.1808 | Ptb. (Wiborg) |
| 8. Charlotte | * 12.7.1744 | Berlin | † 13.2.1780 | Hückelhoven |
| 9. Hermann Friedrich | * 8.5.1747 | Berlin | †12.12.1750 | Berlin |
| 10.Ertmuth Louise | * 13.4.1749 | Berlin | † 9.8.1749 | Berlin |
| 11.Helene Eleonora | * 13.4.1749 | Berlin | † 11.8.1749 | Berlin |
| 12.August Friedrich | * 20.3.1750 | Berlin | † 10.8.1750 | Berlin |
| 13.NN | probably died before baptism or stillbirth | | | |

Prussian army, who however, together with his fiancée, still had to wait almost three years for the wedding: FREDERICK II answered EULER's request of December 2, 1763 in this matter by return mail negatively with the harshly laconic remark that customarily cornets and officer cadets had to put off marriage until they had worked their way up to higher military ranks.[280] This future son-in-law, by the way, tried to persuade EULER to accept a professorship in the Netherlands with a remuneration of 5,000 Gulden, and for a moment, father LEONHARD actually wavered in his decision, already then firmly contemplated, of exchanging Berlin with Petersburg; the country estate in Charlottenburg namely had been sold (if not yet paid for) already in the summer of 1763 for 8,500 Rth., and the very generous invitation of the Empress Catherine II, mediated by Teplov, for relocating to Petersburg was already in his pocket[281], although not yet accepted.

## Break with Frederick — Goodbye Berlin

Answers to the question why Euler eventually had left Berlin after all, have been given many times.[282] They can and will not be described here in detail, but rather discussed according to their weightiness. In the background, there is Frederick's relationship to, resp. disparity with, all that he did not comprehend, and this, most of all, concerns mathematics. This judgment, however, calls for a qualification insofar as the king definitely recognized the usefulness of the mathematical sciences, since otherwise he would not either have made Maupertuis the Academy president, nor would he have tried so hard — though in vain — to secure the famous French mathematician Jean d'Alembert[283] as the latter's successor, nor would he ultimately have accepted Lagrange as Euler's successor. Of course, Frederick II was not able to comprehend the 'French' mathematics of d'Alembert, or even Lagrange, any more than the 'German' one of Euler, and thus the preference for Frenchmen touches on a deeper-seated, psychological problem of Frederick.

Without exception, all Euler biographers see a principal reason for Euler's departure from Berlin in the fact that he, the undisputably most important mathematician of his time, was denied the presidency of the academy by the Prussian king, and this, to be sure, not only because it would have meant a salary increase of at least 50 percent, but because king Frederick, as is known, demanded from his president one other thing, which the upright Swiss could not offer him: worldly polish, French elegance, self-assured manner in the slick courtly society, bubbling, witty salon conversation, and — if possible — aristocratic descent. Euler was certainly more than a little irritated about Frederick already in 1752 thinking of d'Alembert as Maupertuis's successor after the latter had offered the king his resignation as a result of the loss of prestige suffered in the dispute about the principle of least action, and all the more, at least after Maupertuis's demise (1759), did Euler have hopes for the presidency,

JEAN LE ROND D'ALEMBERT. Oil painting by
M. QU. DE LA TOUR

having taken care since 1753 of all the academy business during the latter's absences. However, FREDERICK insisted on D'ALEMBERT, who had become world-famous, not only through his substantial collaboration in DIDEROT's great *Encyclopédie*, but also through his brilliant literary style. As soon as the Seven-Year War was over, FREDERICK II immediately invited D'ALEMBERT to Berlin. The latter promptly came for one summer, but not to take over the presidency of the Berlin Academy, as the king had hoped, but rather to urge him that EULER should be entrusted with this office! This was not entirely expected, for EULER shared — to say the least — over a long time the resentments of DANIEL BERNOULLI toward D'ALEMBERT, as can be learned from various intimate correspondences of EULER with MÜLLER, KARSTEN, and DANIEL BERNOULLI.[284] This changed all of a sudden with the personal acquaintance between D'ALEMBERT and EULER, as the latter's letter to GOLD-BACH from the fall of 1763 shows: *Als sich letztens Mr. D'ALEMBERT einige Zeit hier aufhielt und von Sr. Majestät dem König mit den höchsten Gnaden-bezeugungen überhäufet wurde, hatte ich auch Gelegenheit, denselben persönlich kennenzulernen, nachdem schon seit geraumer Zeit unser Briefwechsel wegen einiger gelehrten Streitigkeiten unterbrochen gewesen, in welche ich mich nicht einlassen wollte. Nun aber ist unsere Freundschaft auf das vollkommenste wieder hergestellet worden; und man kann mir nicht genug beschreiben, mit wie großen*

déduire la loi fondamentale de l'hydrostatique, jusqu'à présent assez mal prouvée par la théorie. Votre théorème sur la route apparente des comètes me paroît très beau, très simple, et très utile; je suis étonné que l'excellent ouvrage que vous avez déjà donné a ce sujet ne soit pas plus répandu, il mérite bien de l'être.

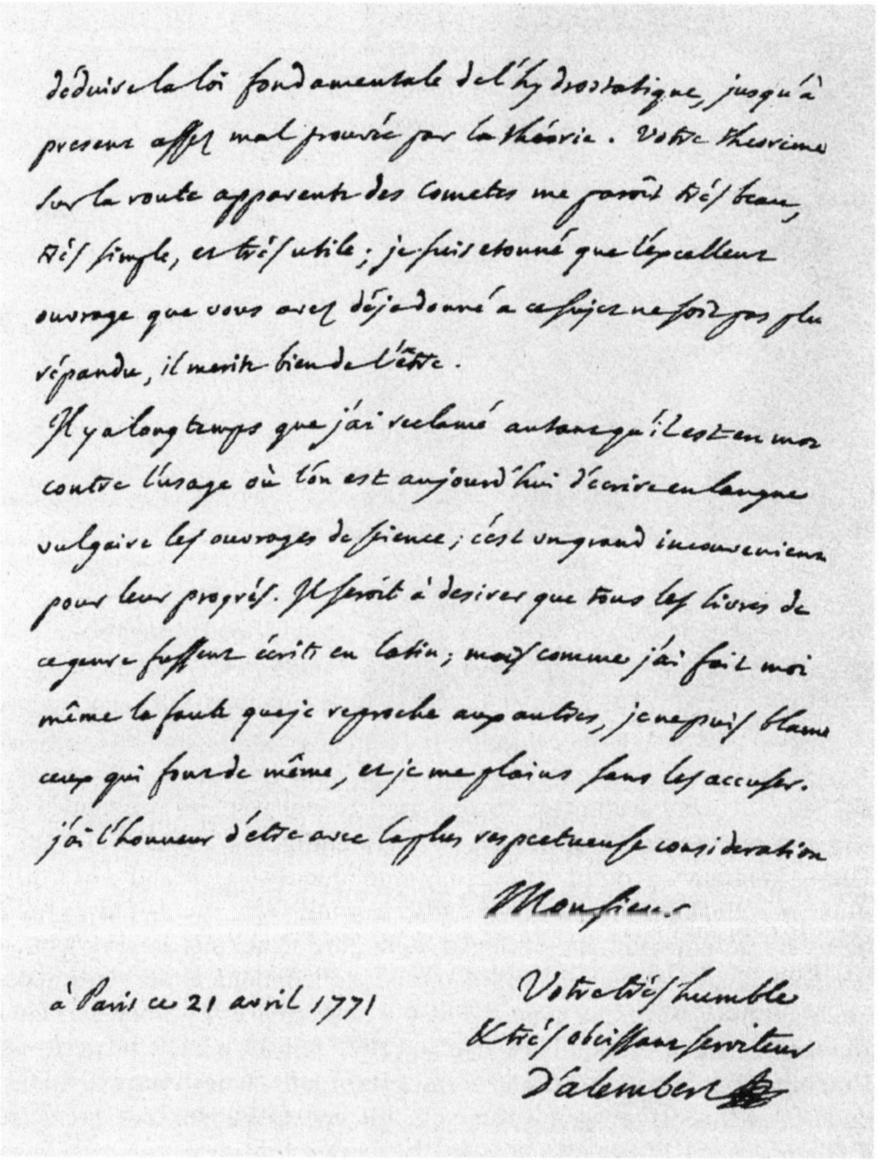

Il y a longtemps que j'ai réclamé autant qu'il est en moi contre l'usage où l'on est aujourd'hui d'écrire en langue vulgaire les ouvrages de science; c'est un grand inconvénient pour leur progrès. Il seroit à desirer que tous les livres de ce genre fussent écrits en latin; mais comme j'ai fait moi même la faute que je reproche aux autres, je ne puis blâmer ceux qui font de même, et je me plains sans les accuser.

j'ai l'honneur d'être avec la plus respectueuse considération

Monsieur

à Paris ce 21 avril 1771

Votre très humble
& très obéissant serviteur
D'Alembert

Last page of a letter of D'ALEMBERT to JOHANN HEINRICH LAMBERT, April 21, 1771

JOHANN HEINRICH LAMBERT. Lithography by
ENGELMANN after a drawing by VIGNERON

*Lobeserhebungen er beständig mit Sr. Königl. Majestät von mir gesprochen. Unter der Hand wird versichert, daß er doch künftigen Mai wieder herkommen und die Präsidentenstelle unserer Akademie antreten würde.*[285] [When Mr. D'ALEMBERT recently spent some time here and was inundated by His Majesty, the king, with the highest displays of favor, I also had the opportunity to meet him personally, after our correspondence had been interrupted already some considerable time ago because of several scholarly disputes in which I didn't want to become involved. Now, however, our friendship has been restored completely; and it can't be put into words with what high praises he continuously was talking about me to His Royal Majesty. Privately it has been affirmed that he will return coming May and will assume the position of president at our academy.][285] Ultimately, D'ALEMBERT did not come to Berlin after all, but preferred, despite the handsome salary approved for him, to remain free in Paris and to work independently and in peace, yet he spared no effort to motivate EULER to remain in Berlin, and to engage himself on his behalf with the king of Prussia.[286] EULER must have heard about this early on after D'ALEMBERT's arrival in Potsdam, for at the end of June 1763, after for a long time he had feared the latter's acceptance of the presidency, he wrote to MÜLLER, astonished: *Monsieur D'ALEMBERT ist schon seit 10 Tagen in Potsdam, soll sich aber rundaus erklärt*

Joseph-Louis Lagrange

*haben, nimmermehr in hiesige Dienste zu tretten und sogar mich zur Praesidentenstelle der hiesigen Akademie vorgeschlagen haben. So gewiß ich aber weiß, daß dieser Vorschlag wird verworfen werden, ebensowenig Neigung habe ich, diese Stelle anzunehmen.*[287] [Monsieur d'Alembert has been in Potsdam for 10 days now, but is said to have outright declared to never ever assume any duties here, and apparently even has proposed me for the presidency of the academy here. But as surely as I know that this proposal will be rejected, as little inclination do I have of accepting this position.][287] And at that it should remain, for Euler still had other reasons.

One of the most compelling was, without doubt, the newly rising influence of the philosophy of Wolffian observance — subscribed to also by the Swiss Johann Georg Sulzer — as well as the Francophilia pushed by king Frederick II, and the favoring of French philosophers of the Enlightenment resp. free-thinkers such as Jaucourt, Toussaint, Castillon, and Helvétius, who for the devout protestant Euler, performing high church functions in the consistory of the French-protestant community[288], made life and work at the academy exceedingly difficult, even unbearable.[289]

Although generally not unforgiving, Euler understandably could not forget that Frederick hardly ever had granted him a personal wish. Let us only be reminded of the royal ban for the marriage of Charlotte Euler to VAN DELEN, of the negative response to EULER's request for granting his brother HEINRICH modest employment as an artist, not to mention EULER's long waiting period until the king granted him, at least partially, the travel reimbursements from Petersburg to Berlin, assured him in 1741, and for which EULER had to beg most humbly. Furthermore, when JOHANN ALBRECHT EULER, after the end of the Seven-Year War, offered to take on the mathematics professorship at the new knight academy to be founded, he received a negative answer on the grounds of still being too young (at 29 years) for this; the professorship then went to the eighteen-year-old son FRÉDÉRIC of JEAN DE CASTILLON, of whom JOHANN ALBRECHT EULER in a letter to KARSTEN sarcastically remarked that he possessed the knowledge of a pupil in secondary school at best.

In the financial affair around DAVID KOEHLER in connection with the calendar sales — it has already been described repeatedly[290] — EULER has undoubtedly put himself in a spot, and the loss of competency resulting from it, combined with the not particularly fortunate activity of the congenial JOHANN HEINRICH LAMBERT[291], has finally put the lid on it.

When EULER had learned that his official offer to return to the Petersburg Academy, communicated on Christmas Eve 1765 to the Russian Lord High Chancellor MIKHAIL WORONZOV, had been accepted, he already on February 2, 1766 requested of king FREDERICK to be released from the Berlin Academy. Since the latter simply did not react, EULER repeated his request two more times. Then, finally, FREDERICK II replied with a short letter of March 17, 1766, in which he acknowledges receipt of EULER's three letters of resignation, and says in his arrogant manner that EULER should withdraw his resignation and not write him anymore about it.[292] The letters of EULER's son JOHANN ALBRECHT to KARSTEN reflect this drama rather impressively:

February 4, 1766: "My father has yesterday applied for leave for himself and his entire family consisting of eighteen souls."

February 15: "My father has requested leave twice already, but to this hour has not received an answer. We live in the most dreadful uncertainty, and are likely to remain in that state for still some time to come."

April 5: "How I would have wished to be able to report to Your Honorable that my father will definitely either stay here, or depart. Unfortunately, we still continue to live in the most dreadful uncertainty. The king persists in his utter silence and will neither let my father leave nor have the good grace of offering him satisfaction or service. My father, however, upholds his Swiss freedom and becomes more and more eager for his departure. He is doing everything in order to be released. He does no longer attend the Academy sessions, he no longer works for the Academy and also does not want to work for it anymore. In one word: he does everything to be dismissed, to be thrown out or, as one says here, to be "kassiert".[293]

In a polite letter of April 30, 1766 EULER in his Helvetic stubbornness insists on the release for himself and his three sons, and by return mail on May 2, there then arrives — under pressure of the Empress CATHERINE II's intervention through the Russian ambassador in Berlin — the laconic royal reply: "With reference to your letter of April 30, I am giving you permission to resign in order to go to Russia. Federic."[294]

Today we know that during this episode DE CATT, the very influential reader of the king, as well as the secret president of the Berlin Academy D'ALEMBERT, out of Paris, had made a concerted effort with FREDERICK for EULER's release in order to avoid a public scandal. As a petty act of revenge of the great FREDERICK he however decreed that EULER's son CHRISTOPH, as his 'military subject', was not allowed to travel with them — the latter had to still wait for more than half a year until he was allowed to enter the Russian army in a higher military rank.

This is how the most famous scholar of his time was dismissed in ingratitude, yet the big loser was the king of Prussia with his Academy. The inestimable loss, however — at least with regard to mathematical research, but not the administration of the Academy — was somewhat compensated for by EULER's successor as director of the Mathematical Class, namely by JOSEPH-LOUIS DE LAGRANGE[295], who also counts among the stars of first magnitude on the mathematical firmament. He was proposed to FREDERICK II by D'ALEMBERT, and the Prussian king acted on it quickly, while at the same time EULER from Petersburg tried in vain to recruit for Russia the truly brilliant, then thirty-year-old LAGRANGE from Turin, with whom he was in correspondence since 1754[296]. The new acquisition was reported by FREDERICK II on July 26, 1766 to D'ALEMBERT in Paris: "Mister DE LA GRANGE shall arrive in Berlin ... and I owe it to your efforts as well as your recommendation that at my academy the one-eyed master of surveying has been replaced by another one who has both his eyes: which will be especially pleasing to the Anatomic Class."

It is more than embarrassing to read how FREDERICK here pokes fun in a most distasteful way at the physical handicap of a man, to whom, incidentally, he owed in an essential way the establishment and functioning of his academy. This also reveals the poor psychological condition of the monarch — and perhaps also his guilty conscience, combined with the anger about his own irreparable misbehavior. In the same letter he writes gloatingly about the shipwreck suffered by EULER and his family during the crossing to the Russian metropolis, and during which EULER's luggage had been damaged: "Mr. EULER, who is extremely fond of the Great and the Little Bear, has moved farther to the north in order to observe them with more leisure. A ship that had loaded up his $x - z$ and his $kk$, has shipwrecked; everything has been lost: This is lamentable, because from it six tomes could have been filled with treatises full of numbers from beginning to end, and now, probably, Europe will be robbed of such pleasant reading."[297]

On June 1, 1766 EULER left Berlin with his entourage — eighteen persons including four servants. His severe bitterness might have been considerably eased during this journey by the unexpected display of goodwill on the part of the higher aristocracy: In Warsaw, STANISLAUS II AUGUST PONIATOWSKI, the last king of Poland (who was to live to see all three partitions of his country), took him in with great warmth, in Mitau the EULER family for four days were the guests of the Duke of Courland, and in Riga, finally, they were granted free lodging along with equipage and two grenadiers as guardsmen. On July 28, the party arrived in Petersburg, where they were given an absolutely triumphant reception.

# 4
## The second Petersburg period 1766–1783

## Firmly in the saddle

In December of 1765, in a letter to the Lord High Chancellor Woronzov transmitted by the Russian envoy Prince Dolgoruki at the Berlin court, Euler formulated his conditions under which he would return to the Petersburg Academy. He demanded, to begin with, the position of a vice president; the president should always be a nobleman at the court, but scientific activities were not to be conducted in a chancellery. For these, a vice president — always a first-rate scientist — ought to be responsible. Secondly, Euler requested a yearly salary of 3,000 Rbl. along with free lodging and heat; thirdly, the exemption of any military quartering; in the fourth place, exemption from duty for his entire luggage, and reimbursement of the travel expenses for fourteen persons (not counting the servants); fifthly, for his eldest son Johann Albrecht the physics Chair with an annual salary of 1,000 Rbl., and in the sixth place, adequate positions for his sons Karl and Christoph at medical resp. military institutions.

Catherine II agreed to these conditions — with the exception of the vice presidency. Since the academy president Count Rasumovski — we remember him from Euler's earlier time in Berlin — was about to embark on a lengthy journey, the empress had actually already appointed a vice president in the person of Count Vladimir Grigoryevich Orlov, the brother of her protégé Grigori Grigoryevich. Also another one of Euler's requests she discretely denied, namely to grant him the title of academic counsel, which he had asked for — presumably to please his family. In doing so, Catherine reasoned, Euler would be put on the same level with a number of people who were beneath him in importance; his fame, according to her, meant more than any title and, finally, his family's esteem was the same as that of the local nobility. On the other hand, she initially gave Euler 10,800 Rbl. for the purchase of a house including furniture, and for the time being made one of her cooks available to them.

Catherine II

Thanks to the good employment of his three sons and the marriage of his two daughters Helene and Charlotte, "the burden of caring for his children, which had always troubled him during his time in Berlin, was taken off Euler's shoulders, and he, carried by princely favor, saw himself entrusted with the reorganization of the scholarly institute whose difficult beginnings he himself had witnessed forty years ago as a humble *élève*. His charge was to bring the academy, which had deteriorated under the earlier government, to new splendor by attracting outstanding personalities. On this subject, he often had lengthy discussions with the empress, lasting many hours. The way he envisioned the reorganization he also laid out in a memorandum[298]."[299]

Thus, Euler was firmly in the saddle again, yet in his personal life he had to endure some hard blows. Already during the last years in Berlin a cataract became apparent in his left eye, and in Petersburg the ailment became rapidly more severe, so that Euler decided on having the cloudy lens surgically removed. The operation, it is true, was successful, but owing to complications he lost his eye, and for the rest of his life, the mathematician was left with only the tiniest residual vision[300]. The fire disaster of 1771 has already been mentioned. To alleviate its effects for the Euler

LEONHARD EULER's house in St. Petersburg from 1766 to 1783. Originally, the house had two floors, the top floor was added in the 19th century.

family, the empress donated the amount of 6,000 Rbl. for the construction of a new house which still today stands near the academy building in St. Petersburg — with a memorial plaque attached to it. On November 10, 1773 EULER's wife KATHARINA dies. Three years later, EULER, wanting to be self-sufficient and not dependent on his children, marries her half-sister SALOME ABIGAIL GSELL, following dramatic entanglements — we will return to this later.

A rather impressive description of EULER's condition at the time we owe to JOHANN III BERNOULLI, a grandson of JOHANN I BERNOULLI, who as astronomer in Berlin was EULER's colleague at the Academy (and who today is known almost exclusively only through his multi-volume "travel reports"). He visited EULER in the summer of 1778 and writes in connection with the latter's eleventh note book, which he was able to look through in Petersburg: "His health is still quite good, which he owes to a very moderate and regular life style; and he is able to make use of his vision, for the most part lost since long ago, and for some time entirely, still even better than

A sheet of sketches by Leonhard Euler's son Johann Albrecht, presumably from the time between 1766 and 1770 in St. Petersburg. The two awkwardly scribbled human figures probably represent the Euler pair of parents. If the presumption were true, we would have here the only surviving portrait of Katharina Euler.

JOHANN III BERNOULLI

many realize: It is true that he cannot recognize people by their faces, nor read black on white, nor write with pen on paper; yet with chalk he writes his mathematical calculations on a blackboard very clearly and in rather normal size; these are immediately copied by one of his adjuncts, Mister Fuss and Golovin (most often the former) into a large book, and from these materials are later composed memoirs under his direction."[301]

From the (unfortunately largely unpublished) letters of JOHANN AL-BRECHT EULER to his uncle SAMUEL FORMEY, who as secretary of the Berlin Academy was his fellow-member and colleague, we know that LEONHARD EULER's harmonic relationship with the Prussian king and his Academy was soon restored, more or less. In 1767, the Russian empress — surely not lastly because of political motives — was appointed by FREDERICK II as honorary member of the Berlin Academy, and barely a year later as a regular foreign member.[302] CATHERINE II, in turn — favorably influenced probably also by EULER, but surely by his son JOHANN ALBRECHT — had FREDERICK elected as an honorary foreign member of the Petersburg Academy. It is true, though, that this was preceded by a rather lengthy visit to Petersburg by the royal brother, Prince HEINRICH, in the summer until the fall of 1770[303], during which the waves between EULER and FREDERICK II apparently seem to have been smoothed. At least so it appears from the last six letters these so disparate partners have still exchanged for a year after EULER had resumed correspondence with the king of Prussia since the latter's laconic permission of May 1766 for him to leave the country[304], and this only in connection with a new work of EULER (E. 473, 473A) on the establishment of pension funds and the calculation of pensions. Here, once again, EULER's conciliatory character manifests itself.

## The "Algebra for beginners"

During the second Petersburg period, EULER produced more than 400 papers on the most varied topics in pure and applied mathematics, astronomy, and physics. About important principal works, actually printed in Petersburg but, in part, already written in Berlin, has already been reported. We have in mind EULER's books on the theory of ships, on optics, the monumental integral calculus, and the *philosophical letters*. (With astronomy we will deal in the next section.)

Most likely still in his Berlin period, EULER set out to write his probably most popular mathematical book, his two-volume opus *Vollständige Anleitung zur Algebra*[305] [Complete guide to algebra[305]], which he dictated to his assistant, a mathematically completely unencumbered former tailor, who — allegedly a mediocre fellow — was then to have understood it all. It is a legend, initiated by an editorial "pre-report", that EULER wrote resp. dictated to his assistant the *Algebra* immediately after his loss of sight in Petersburg for the purpose of selfcontrol, because firstly, EULER absolutely did not need such "selfcontrol", and secondly, he became almost completely blind, as is well known, only after the cataract operation of 1771, when the book had long been out in several editions. Thirdly, in the text of the first volume, there are a few passages which can, if not must, be interpreted as indicating a date 1765/66 of composition[306].

This work — especially fascinating in view of EULER's masterly didactic skill — truly became a bestseller. It appeared 1768/69 first in Russian translation, then 1770 in the original German version, 1774 in the French translation of JOHANN III BERNOULLI, and finally in the English and Dutch language in many editions. The *Algebra*, as the book is called for short, introduces an absolute beginner step by step from the natural numbers via

Frontispiece of EULER's "Algebra", Petersburg 1770

the arithmetic and algebraic principles and practices through the elementary theory of equations right to the most subtle details of indeterminate analysis (Diophantine equations); it still is — in the judgment of today's foremost mathematicians — the best introduction into the realm of algebra for a "mathematical infant". With good reason, the "great EULER edition" of 1911 was inaugurated with this volume. The great mathematician LAGRANGE, EULER's successor at the Berlin Academy, did not consider it below his dignity to provide the work in the first French edition with valuable additions. They fill there 300 pages, and in the first volume of EULER's *Opera omnia* they were reprinted in the original French language.

In the German-speaking areas, the *Algebra* experienced the widest circulation through its inclusion in Reclam's Universal Library, where it figures as the only mathematical book and was printed from 1883 till 1942 in 108,000 copies.[307] There is only one other book in the field of mathematics which in the entire history of culture has had a comparable success in sales: the "Elements" of Euclid, after the Bible the most frequently printed book of all.

# Astronomy*

EULER's works on astronomy[308] exhibit a wide spectrum: The determination of planetary and cometary orbits on the basis of a few observations, methods for calculating the parallax of the sun, the theory of atmospheric refraction of rays without which spherical astronomy could never have been put on a firm basis, alternate with considerations about the physical nature of comets and also about the slowing down of planetary motion due to the (hypothetical) resistance of the ether. His most important memoirs, with which he garnered an impressive number of prizes from the Paris Academy, relate directly or indirectly to celestial mechanics, the theoretical astronomy, then of particular interest — a branch of science which demanded the highest efforts of the greatest mathematicians of the time.

Already soon after NEWTON's death, it was noticed that the observed orbits — particularly of Jupiter, Saturn, and the moon — deviated significantly from the values calculated according to NEWTON's gravitational theory. For example, the respective calculations of the perigee of the moon by CLAIRAUT and D'ALEMBERT (1745) yielded a period of eigtheen years, whereas observation yielded one of only nine years. This unpleasant state of affairs called in question NEWTON's theory as a whole. For a long time, EULER — and not he alone — was of the opinion that NEWTON's law of gravitation required several corrections. CLAIRAUT, however, tried to explain the discrepancy between theory and practice by the fact that up to now the corresponding differential equation had been accounted for only in a first approximation, and he thought (rightly so) that consideration of also the second approximation would offset the difference. EULER at first did not agree, but (from Berlin) had the Petersburg Academy announce a corresponding prize question. Subsequently, he realized that CLAIRAUT was correct after all, and recommended him for the prize, which the Frenchman indeed received in 1752. Still, EULER — stubborn as always — was not quite satisfied. He wrote his first *lunar theory*[309], in which he developed a fundamental method for the approximate solution of the *three-body problem*. With this problem — the question here is to describe mathematically the motion of three masses (thought of as points) subject to gravitational forces — EULER came into contact already quite early, still in the period of Basel. He soon realized the immense difficulties of solving the problem (a

CLAUDE ALEXIS CLAIRAUT

general solution is impossible), and first looked at special cases, which he could indeed solve by means of a method which today is called "regularization". In a memoir[310] submitted in 1747, he formalized the restricted three-body problem (*problème restreint*), usually ascribed to CARL GUSTAV JACOB JACOBI and HENRI POINCARÉ, and thus was the first to tackle analytically the general problem of perturbation.

EULER's first *lunar theory*, by the way, had a practical consequence not to be underestimated: The astronomer TOBIAS MAYER[311] in Göttingen used EULER's formulae to produce lunar tables which made it possible to determine the position of the earth satellite, and with it, the geographic longitude of a ship at high sea, with an accuracy which, in navigation, up to that time had never been attained. The British parliament in 1714 had offered a considerable monetary prize for the determination of the longitude at high sea to within an error margin of one half degree. This prize was awarded in 1765 for the first time: MAYER's widow received 3,000 pounds, and EULER 300 pounds for the theory underlying MAYER's tables. These lunar tables became part of all navigational almanacs and served merchant shipping for more than a century.

In the years 1770 to 1772, EULER worked out his second *lunar theory*[312] whose advantages over the first one, however, could be properly appreciated only after the development of the magnificent ideas of GEORGE WILLIAM HILL[313], which in turn required EULER's lunar theory as a stepping stone. When in this context EULER, overestimating the scope of his methods, at one time also announced an alleged solution of the three-body problem – and this in a prize-winning memoir! – one has to leniently consider that at that time the impossibility of a solution could not yet be proved. EULER, also for the science of the stars, turns out to be a star of the first magnitude.

G. K. Mikhailov[314]

# The discord about Leonhard Euler's second marriage

An ordinary episode from the 18th century,
reconstructed after the testimony of a participant

LEONHARD EULER's private life during the seventeen years of his second abode in Petersburg is thoroughly documented in two voluminous correspondences. EULER's eldest son, JOHANN ALBRECHT, secretary of the Petersburg Academy since 1769, was in regular correspondence through all these years with SAMUEL FORMEY in Berlin, who was his uncle-in-law, as it were, since FORMEY's first wife, who died at an early age, was a sister of JOHANN ALBRECHT's mother-in-law, who passed away at even an earlier age. This correspondence consists of several hundred letters – often very detailed – with excerpts from the systematically kept diaries of J. A. EULER, and is well preserved – apart from some (important) gaps. The second correspondence of interest to us here was, since 1773, between LEONHARD EULER's private secretary, NIKLAUS FUSS, also a member of the Petersburg Academy and a grandson-in-law of LEONHARD EULER (he married a daughter of J. A. EULER), and his father in Basel. Unfortunately, we are in possession only of copies of a few excerpts of this large correspondence. These two collections, which have preserved for us many biographical details from the life of the great scholar, to this day have never been investigated thoroughly. Here, we merely want to shed some light on a small aspect of EULER's family life, as gained from the letters of J. A. EULER.

EULER's first marriage with KATHARINA GSELL has already been reported on. With this brave woman he lived together almost forty years and left all the everyday domestic worries entirely to her. She bore him thirteen children, of whom eight died at the infant or childhood age; the other five, however, grew up to adulthood and later delighted him with many grandchildren. In the seventies, four of his children lived in Petersburg or nearby. The eldest son, JOHANN ALBRECHT, even lived in the house of his father, where he occupied with his own family the ground floor, and worked there as his closest collaborator. EULER's second son, KARL, was a court physician, and the youngest son, CHRISTOPH, was an officer in the Russian army. The eldest daughter, KATHARINA, married the formerly Prussian, and later Russian, officer CARL VON BELL; only EULER's younger daughter, the Baroness CHARLOTTE VAN DELEN, left Russia and moved in the summer of 1769 with her husband to the latter's country estate in Jülich at the lower Rhine.

The daily routine of family life of the great scholar was suddenly interrupted by the death of his caring wife. On Leonhard day, Wednesday, November 6, 1773, she didn't feel well, became bedridden, and the following Sunday already, early in the morning, the whole family had to bid farewell to her; fully conscious, she blessed her family, giving them some well-meaning admonitions, and passed away on November 10, 1773 (old style) at half-past eleven at the age of 66. At the solemn funeral on Wednesday, November 13, close to a hundred mourners paid their last respect to the deceased. (More than thirty coaches are said to have driven up, and the expenses of the funeral — according to the indications of the exact JOHANN ALBRECHT — are said to have exceeded 500 Rbl.)

LEONHARD EULER, who previously probably never had anything seriously to do with the household, was now forced to look around for help. But the aging scholar wanted to remain independent and not link his way of life with that of his son's family living in his house, which is the reason why he pondered the question of a possible second marriage. EULER knew well that this idea could not meet with unreserved enthusiasm on the part of his children, and therefore decided — without making a great commotion about it — to look around for a modest woman, relatively advanced in years. Since, because of the dimness of his eyesight (almost blindness), he hardly ever used to pay social visits anymore, he had to seek his intended wife among the very narrow circle of visitors to his house.

In the fall of 1775, EULER directed his attention to a certain Frau MET-ZEN. She was repeatedly widowed, had served for a while as the housekeeper for General BAUER in Petersburg, and — as later emphatically stressed by JOHANN ALBRECHT — was of "low descent". Frau METZEN, about two years earlier, had been introduced to the social circle of EULER and his children by an old family acquaintance, a certain Frau MÜLLER (whose affairs had been managed for a long time by JOHANN ALBRECHT), and she often appeared for meals.

At one time during the Christmas holidays, LEONHARD EULER informed his (long grown-up) children all of a sudden that he had decided to marry Frau METZEN, and also that he already had given her his promise. This announcement the father made to his children on the street — on the way back from church — whereupon they reacted with shock and indignation. According to JOHANN ALBRECHT's words, his brothers are to have flown into a rage, and the women burst into tears. The children immediately took the field with all sorts of possible and impossible arguments in order to convince their father of the irrationality of his decision. The purport was that the Petersburg society (and even the empress!) would not understand such a marriage. The fact of the matter, however, was primarily the fear of the children of possible losses in the impending disposition of property by testament. The father EULER remained unrelenting, and the consequence was a family split. After this, the children used all means at their disposal and tried to force the poor Frau METZEN herself to renounce the proposed marriage. In the course of these events, EULER was taken ill with a severe fever, but his son — the physician — brought him back to health within two weeks, whereupon peace seemed to have returned to the family.

But basically, the father was unyielding. Half a year later, on Wednesday, July 20, 1776 (old style), EULER categorically declared to his children that he was now going to marry the half-sister of his first wife, SALOME ABIGAIL GSELL. This time, any objection was ruled out, and the children had to give in. SALOME ABIGAIL — a younger daughter of GEORG GSELL and his third wife MARIE, née GRAFF — was born on July 6, 1723 (old style) in Russia, received a Russian education, spoke a broken German, and at the time was 53 years old. What may have calmed down EULER's children

Portrait of Euler in his old age. Oil painting by J. Darbes, 1778

was the fact that their future stepmother was no foreigner, and that it was quite improbable to obtain new siblings from her! She was a small woman, inconspicuous, humble, not particularly pretty, but probably very home-loving, who presumably had lived already previously in EULER's house.

The formalities were quickly settled: On Sunday, there was an announcement in church; the day before, EULER informed his children about his testament, and the following Thursday, July 28, the wedding took place at home.

The testament set off new emotions in the large Euler family; the children were highly dissatisfied with it. The eldest son JOHANN ALBRECHT obtained separately, first of all, one twelfth of the house, in addition to his share of the remaining eleven twelfths in silverware (an inventory thereof was taken) and in the library. The rest — furniture, carriage, mirrors, clocks, etc. — should pass on to the widow after EULER's death. In addition, some questions of pension were put in order: The children, during the father's lifetime, received each 200 Rbl. annually and free housing, and to the widow they were to pay a yearly rent of 200 Rbl.

JOHANN ALBRECHT felt that, as co-owner of the house and direct assistant to the old father, he should be allowed to lay claim to a larger share, while the other children considered it unjust that he should receive more than they did. It is true that as time went on, the relationship between the children returned to normal, but during the reception in the house after the wedding, there was a tense atmosphere, as JOHANN ALBRECHT later reported.

The wedding itself, as is fitting, was quite ceremonial. The invited guests gathered in EULER's home toward five o'clock in the evening. Altogether, there were about thirty persons; the grandchildren were dressed in festive clothes. The loyal NIKLAUS FUSS was master of ceremonies, and after six o'clock the pastor arrived. EULER retired with the pastor, his bride, and an acquaintance — the artillery captain HAECKS as witness — to his study, where they all signed EULER's last will and testament. This document and the silverware inventory were given for safekeeping to JOHANN ALBRECHT as the eldest son, and the bride received a copy prepared by EULER. In advance, EULER had cleared the way for securing the assistance of the imperial court with regard to the new matrimony. Consequently, after the papers were signed, he could pass on to the empress the arrangement, according to which the pension of 1,000 Rbl., awarded to his first wife, in case of his death, was to be transferred to his second wife. Then the wedding proper began. The pastor delivered a speech appropriate for the occasion; the children, relatives, and guests congratulated the newlyweds, and the day ended to everyone's contentment with a festive supper.

"This step creates much sensation all around" Niklaus Fuss commented on the second marriage of his teacher. "Fools laugh at him. Sensible people regret that a wise man in his final years does something foolish, and if they know his domestic conditions, they excuse a half-blind man who is wise enough not wanting to be dependent on his children." By the way, the children were decent enough, after Euler's marriage, to express the full respect to their stepmother. Leonhard Euler still lived peacefully for seven years with his second wife. She outlived him by eleven years and died in Petersburg on December 25, 1794.

# The end

In Petersburg, Euler now officially obtained the position, which he had already held during the decades in Berlin: *Spiritus rector* of the scientific operation at the Petersburg Academy. Besides, he took on — as already in his first Petersburg period — the function of an expert and consultant in practical technical problems, as for example in the establishment of a widow's fund (1769), in the assessment of the project of a 300 meter long suspension bridge over the Neva by Ivan Kulibin (1776), and in the determination of the incline and speed of the Neva current (1780).

Being the oldest member of the Academy, Euler also had to preside over the sessions of the Academy, the "conference", which his son Johann Albrecht, since 1769, as the secretary had to protocol. In addition, he was a prominent member of the new managing organ of the Academy, a commission reporting only to the director. While the president, Rasumovski, cared rather little about the Academy, the directors, on the other hand, changed frequently. With Orlov and his occasional substitute Alexey Reshevski, the two Eulers got along relatively well, but under the directorate (1775–1782) of Sergey Domashnev, a sort of Schumacher in type and behavior,

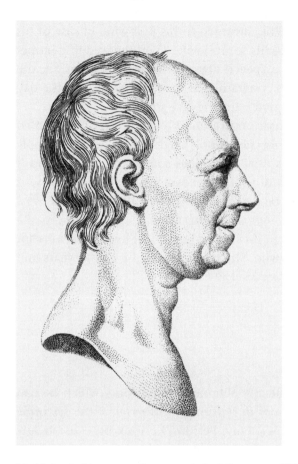

Marble bust of LEONHARD EULER, made by I. RACHETTE, 1784

they — and others as well — had a difficult time. This led to the resignation of both EULERs from the commission, to which they belonged since
1766, and to LEONHARD EULER keeping more and more away from the con-
ferences. Empress CATHERINE eventually had to dismiss DOMASHNEV; she
replaced him by her intimate friend, the still young Countess YEKATERINA
ROMANOVNA DASHKOVA. About this very remarkable lady, STIEDA at the
time wrote: "Animated with the best intentions, equipped with sufficient
education and intelligence to do justice to the difficult position, she was
generally on good terms with the academicians. For the importance of the
EULERs, she had a good eye."[315]

This is fully and entirely correct, and it agrees with the two pages which DASHKOVA in her memoirs dedicated to the old LEONHARD EULER. Therein she describes how she let EULER accompany her in January of 1783 to her first — and probably his last — Academy session: As the Professor and Privy Councillor JAKOB STÄHLIN, with foregone conclusion, claimed the honorary seat next to the director's chair, she turned to EULER with the words: "Please be seated wherever you want, and the seat you choose will of course be the first of all the others."[316]

About EULER's last days of his life, and the details of his death, we are rather well informed through a few of his contemporaries and family members who were present (J. A. EULER, FUSS, ABEL BURYA[317], CONDORCET). During the first days of September, he occasionally felt a bit dizzy, which however did not deter him from occupying himself with the mathematical foundations of the sensational balloon flight, on June 5, 1783, of the brothers MONTGOLFIER in Paris. The formulae sketched by EULER — they were found on the day of his death on two of his large writing slates — form the skeleton of his last work. It was written out by EULER's son JOHANN AL-BRECHT (in Latin) and immediately sent to the *Académie des Sciences* in Paris, where it appeared — with an *Avertissement* in French — in the volume of the *Mémoires* published in 1784.[318]

The morning of September (7)18, 1783, EULER spent in the usual way. He gave a lesson to a grandchild, and during lunch discussed with his assistants FUSS and LEXELL the orbit of the planet Uranus discovered on March 13, 1781 by HERSCHEL. A bit later, toward five o'clock, he went to the grandchild again to joke with him a little and to drink a few cups of tea. Sitting on the couch, he smoked a pipe, which suddenly slid from his hand. *Meine Pfeife!* [my pipe!] EULER is said to have yelled loudly; he bent forward to pick it up, but without success. Then he grabbed his forehead with both hands and lost consciousness with the cry *Ich sterbe!* [I am dying!]. EULER suffered a stroke and did not regain consciousness. The agony continued until about eleven o'clock at night, then EULER "had stopped to calculate and to live"[319].

He passed away at the age of 76 years, five months, and three days.

The death of EULER was perceived in the scientific world as a great loss, and the important academies paid the utmost respect to their best whom they had lost — first among them, of course, the Petersburg Academy, which

LEONHARD EULER
1707–1783
MATHEMATIKER, PHYSIKER,
INGENIEUR, ASTRONOM UND
PHILOSOPH, VERBRACHTE IN
RIEHEN SEINE JUGENDJAHRE.
ER WAR EIN GROSSER GELEHR-
TER UND EIN GÜTIGER
MENSCH.

Commemorative plaque for LEONHARD EULER at the "Klösterli" in Riehen, Kirchgasse 8. It was made in bronze by ROSA BRATTELER and solemny unveiled in 1960 at the occasion of the 500-year-jubilee of the University of Basel. The text of the plaque is due to OTTO SPIESS. The portrait relief follows the plaster relief of D. RACHETTE, 1781.

in 1784 not only dedicated to him a — incidentally very true-to-life — marble bust, but in 1837, after one had again, at the occasion in 1830 of the burial of one of his daughters-in-law, come across his grave on the Lutheran Smolenski cemetery on the Vasilyevski island, erected a simple, but impressive, monument in granite with the plain inscription

LEONARDO EULERO
ACADEMIA PETROPOLITANA
MDCCCXXXVII

At the occasion of the 250th anniversary of EULER's birth, this commemorative stone, together with the remains of the great mathematician, was moved to the old St. Lazarus cemetery at the Alexandre-Nevski monastery in St. Petersburg and placed near the grave of LOMONOSOV.[320]

Hardly a round anniversary of the birth or death of EULER has passed by without some ceremonial or commemorative volume having appeared in his honor, but the most appropriate memorial for LEONHARD EULER, without doubt, is the edition of his collected works through public and private Swiss institutions, which was begun in the jubilee year of 1907 and still has not yet been completely finished.[321]

# Epilogue

EULER's prestige and influence were impressive already in his lifetime. For about two decades he was (according to ANDREAS SPEISER) the spiritual leader of the educated circles in the protestant part of Germany. He rendered invaluable services as "golden bridge between two academies" (WINTER), of which his correspondence forms an equally clear testimony as the fact that during his Berlin period there are 109 publications in the "Petersburg Commentaries" written by him, as opposed to 119 in the *Mémoirs* of the Berlin Academy. Indeed, EULER definitely had enough stamina to work full-time at both academies, and neither of them alone could have published all of his writings and contributions; even both together did not have an easy time handling the sheer inexhaustible flood of his production. Purely from the point of view of work performance, EULER does not rank behind the most productive exponents of mankind, as for example VOLTAIRE, GOETHE, LEIBNIZ or TELEMANN. We reproduce here a tabular survey (prepared by ADOLF PAVLOVICH YUSHKEVICH), ordered by decades, regarding the quantity of writings made ready for the press by EULER himself — without, to be sure, taking into account a few dozen works which could not yet be dated:

| years | works | % | years | works | % |
|-------|-------|-----|-----------|-------|-----|
| 1725–1734 | 35 | 4 | 1755–1764 | 110 | 14 |
| 1735–1744 | 50 | 10 | 1765–1774 | 145 | 18 |
| 1745–1754 | 150 | 19 | 1775–1783 | 270 | 34 |

With regard to special topics, the respective shares in percentages look about like this:

| | |
|---|---|
| algebra, number theory, analysis | 40% |
| mechanics and the rest of physics | 28% |
| geometry, including trigonometry | 18% |
| astronomy | 11% |

| naval science, architecture, ballistics | 2% |
|---|---|
| philosophy, music theory, theology and what is not included above | 1% |

This listing does not include either the ca. 3,000 pieces of correspondence known so far, or the unedited manuscripts. The distribution of the purely mathematical works of EULER can be seen from the following listing:

| algebra, combinatorics and theory of probability | 10% |
|---|---|
| number theory | 13% |
| foundations of analysis and differential calculus | 7% |
| integral calculus | 20% |
| infinite series | 13% |
| differential equations | 13% |
| calculus of variations | 7% |
| geometry, including differential geometry | 17% |

Altogether, EULER has won twelve international academy prizes, not counting the eight prizes of his sons JOHANN ALBRECHT (7) and KARL (1), which essentially, without scruples, could be booked on his account. The French king LOUIS XVI presented EULER with 1,000 Rbl. for his "second theory of ships", and Empress CATHERINE II, who did not want to be outdone, then bestowed upon him twice this amount, so that in 1773 the Petersburg old master could pocket an additional yearly salary.

Unanimous is the judgment of the most important mathematicians after EULER. LAPLACE used to say to his students: "Read EULER, read EULER, he is the master of us all!", and GAUSS declared concisely: "Studying the works of EULER remains the best school in the various fields of mathematics and cannot be substituted by anything else." Indeed, through his books, which are always distinguished by the greatest efforts toward clarity and simplicity and represent the first textbooks in the modern sense, EULER became not only the teacher of the Europe of his time, but so remained far into the 19th century: The works of BERNHARD RIEMANN, one of the most important representatives of the *Ars inveniendi* in the grandest style, show unmistakable EULERian traits. ABRAHAM GOTTHELF KÄSTNER, to whom we owe the first German history of mathematics of note, coined the excellent comparison that, with regard to mathematical style, D'ALEMBERT was the German,

The EULER medal of the Moscow Academy of Sciences, minted in 1957. It was occasionally awarded to scholars who rendered outstanding services to the research on EULER.

and EULER the French, and C. G. J. JACOBI agreed with this assessment. HENRI POINCARÉ reports that (according to THEODORE STRONG) EULER was the God of mathematics, whose death marked the decline of the mathematical sciences. And indeed, EULER, D'ALEMBERT and LAGRANGE, who in the last third of their century formed something of a triumvirate, undeniably had the feeling of an impending decadence, as can be seen from their correspondence. If they believed to have no spiritual heirs, then this is probably related to "on the summits one is alone".

But also other prominent contemporaries seemed to have had similar feelings. Thus, DIDEROT, the head of the *Encyclopédie* (who, incidentally, knew more about mathematics than is generally assumed), wrote in 1754: "We are facing a great upheaval in the sciences. Given the tendency of today's minds to, what seems to me, are morals, belles-lettres, natural history, and experimental physics, I would almost ascertain that in Europe, before the lapse of a century, one will not be able to count three great mathematicians. This science will suddenly remain at the place where the BERNOULLIS, EULER, MAUPERTUIS, CLAIRAUT, FONTAINE, D'ALEMBERT, and LAGRANGE have left it. They will have erected the columns of Hercules. It is impossible to surpass them. Their works will endure in the centuries to come like those Egyptian pyramids, whose stone masses covered with hieroglyphs evoke in us a frightening idea of the power and the resources of the people who had built them."[322]

Well, history has soundly disproved such resentments, since nowhere more fittingly than in the realm of mathematics are valid the words of JOHANNES: "The spirit goes where it wants to go."

Frequently, into our days — usually unjustly — one hears about alleged definite weaknesses in the work of EULER, mainly about the presumed inadmissible dealings with the concept of infinity, be it in the large (theory of series) or also in the small. But EULER could not possibly have concerned himself with criteria of convergence and continuity in the modern sense, as also with the logically precise and rigorous foundation of analysis in the sense of the *Ars demonstrandi* of a CAUCHY, BOLZANO, or WEIERSTRASS, since a rigorous proof, say of CAUCHY's criterion of convergence, was made possible only after the definition of real numbers — thus 1870 at the earliest. He relied — unsuccessfully only in very isolated instances — on his remarkably sure intuition and algorithmic power. And does not precisely EULER, who has investigated and researched more than any other mortal, have an unconditional claim to the word of KARL WEIERSTRASS, the master of rigor: "It is self-evident that to a researcher, as long as he is investigating, every means is allowed", this all the more as GEORG CANTOR, the creator of set theory, sees the nature of mathematics precisely in its freedom? Certainly, EULER's analytic-algorithmic concept of a function — a Bernoullian inheritance — is too narrow and special and actually demands obvious, but from today's perspective prohibited, generalizations whose dangerous obstacles EULER was able to circumvent only because of his boundless imagination — *Conditio sine qua non* for a creative mathematician — and an almost incredible algorithmic virtuosity, which permitted him to attack the problems at hand from many different angles, and to check, and if necessary, correct the results obtained. In this regard, the most effective vindication of EULER in our time we owe to the Darmstadt mathematician DETLEF LAUGWITZ, the true founder of the so-called nonstandard analysis resp. infinitesimal calculus, which he has developed jointly with CARL SCHMIEDEN.[323]

ANDREAS SPEISER, who dedicated a large part of his life to EULER's work, emphasized to his students over and over again: "There are still many treasures to be lifted from EULER's work, and those who want to chase priorities cannot find more fertile grounds." Indeed, some time will still have to pass until the enormous opus will be fully available in print, and a definitive work biography of the most prominent Swiss abroad is not yet in hand. Understandably so — such an undertaking would be synonymous with writing a universal history of the mathematical sciences of the entire 18th century.

# Translators' Notes

Citations from original sources — mostly letters — written in German (in the orthography of the time) are reproduced here as is if the authors are one of the principal players, EULER and the BERNOULLIS, with the English translation immediately following within brackets. Otherwise, and also in cases of citations which have previously been translated into German, only the English translation is provided.

A few factual changes which have occurred, or come to light, since the publication of the original text have been incorporated in this translation without special mention.

Transliteration of Cyrillic characters follow the system used by the Library of Congress (see, e. g., A. J. LOHWATER, *Russian-English Dictionary of the Mathematical Sciences*, American Mathematical Society, Providence, RI, 1961, p. 1).

# Notes

Primary and secondary literature are cited in text and notes according to the following rules:

1. W o r k s of Leonhard Euler are quoted in the text normally with an abbreviated title (mostly translated into English), in the Notes according to the work edition included in the Bibliography of this book, which is arranged in four series, and of which, as of today, there exist 76 quarto volumes. Euler's works are numbered according to the bibliography of G. Eneström, to which we adhere, and are identified with the respective number, placed in front of the location of a work. For *Opera omnia* we use O., the series numbers are given in roman, and the volume numbers in arabic numerals. Example: E. 65: O.I,24 refers to Euler's work *Methodus inveniendi lineas curvas* ... in Volume 24 of the first series, which carries the Eneström number 65. In each of the four volumes of the fourth series that have appeared so far, there is a table, with the help of which any work of Euler with known Eneström number can easily be found in the *Opera*.

2. L e t t e r s of and to Leonhard Euler are identified by their Résumé number in the first volume of Series IVA, be they published or not. Example: O.IVA, 1, R 465 refers to Euler's letter to G. Cramer of July 6, 1745 (on p. 93 of the first volume of the Series IVA). If the letter partner is named in the text, normally only the date and Résumé number of the letter is given within round parentheses. Under the Résumé number in O.IVA,1 one can find information about the respective letter as well as a brief indication of its content.

3. S e c o n d a r y   l i t e r a t u r about Euler are sometimes linked to the "Burckhardt-Verzeichnis" (BV) in the Basel memorial volume (EGB 83, p. 511–552). This contains ca. 800 work titles in altogether fourteen languages. Example: BV Spiess (1929) refers to O. Spiess: *Leonhard Euler*, Frauenfeld, Leipzig 1929. − Verdun-Bibliography.

4. O t h e r   f r e q u e n t l y   u s e d   a b b r e v i a t i o n s like DSB, IMI, *Harnack, Euler's correspondence, Euler-Goldbach* can easily be inferred via the Bibliography.

[1]    EGB 83
[2]    Cf. Bibliography
[3]    Original in the Archive of the Russian Academy of Sciences in St. Petersburg, where there is also the bulk of Euler's manuscripts. We thank the directorate of the AAN St. Petersburg for the release of the photographs of the originals and for permission to publish them.
[4]    K. Euler (1955)
[5]    *Razvitiye idey* (1988)
[6]    Raith (1983)
[7]    Staehelin (1960)
[8]    Raith (1983), p. 460

Notes

9   Jakob Bernoulli: *Opera*. Geneva 1744, p. 361–373. Its content, on the whole, is elementary and offers almost no mathematical insights new at the time.

10  More details in Raith (1983), p. 468, Note 5

11  Raith (1983), p. 468, Note 9

12  About life and activity of Paulus Euler, especially about his attitude to pietism, see Raith (1983) and (1979)

13  Cf. Raith (1980), p. 102

14  Herzog (1780), p. 32f.

15  Cf. DSB XI, article by K. Vogel

16  Burckhardt-Biedermann (1889)

17  *Magni Euleri praeceptor in mathematicis*. Letter from Daniel Bernoulli to Leonhard Euler of Sept. 4, 1743 (O.IVA,1,R 151)

18  Schafheitlin (1922)

19  Schafheitlin (1924)

20  About the role of a "respondent" in such disputes, cf. Staehelin (1957), p. 135f.

21  Cf. Wolf (1862), p. 88, footnote 3

22  Unfortunately, the text of the lecture has not been handed down to us. About this uncertainty, cf. Aiton (1972)

23  Börner (1752), p. 99f.

24  J. Bernoulli (1742), Vol. 2, §26, p. 615f. (Transl. from the Latin by O. Spiess)

25  *Doctissimo atque ingeniosissimo Viro Juveni Leonhardo Eulero* (R 191)

26  *Clarissimo atque doctissimo Viro Leonhardo Eulero* (R 194)

27  *Viro Clarissimo ac Mathematico longe acutissimo Leonhardo Eulero* (R 201)

28  *Viro incomparabili Leonhardo Eulero Mathematicorum Principi* (R 226)

29  Such prize questions were posed by the big academies in the 18th century – the Paris, Petersburg, and Berlin Academy – almost regularly and constituted an essential stimulant of research. Their solutions were always submitted anonymously in the form of memoirs or even books, which, if they won a prize, were also printed at the expense of the respective academy.

30  Condorcet (1783), p. 288 (in: O.III,12)

31  The respective repudiation of this statement as a legend in EGB 83, p. 81, Note 11, has turned out to be an error, as was pointed out by K.-R. Biermann (Berlin) in two letters to the author of August 12 and 17, 1988.

32  Transl. from the Latin by O. Spiess

33  All the same, one can prove that Euler has also taken up experiments. Mikhailov (1959), p. 258f.

34  About the procedure of electing a professor in Basel, cf. Staehelin (1957)

35  About the building history of St. Petersburg, of the fortress island Kronstadt, and of the navy and mercantile marine, as also about the sociological and demographic conditions, cf. Donnert, p. 10; Buganov, p. 199, p. 291–338, p. 402f.

36  Finster/van den Heuvel (1990), p. 45f.

37  Buganov, p. 368–373

38  Grau, *passim*

39  Grau, p. 120

40  Ostrovityanov, p. 30f.

41  Spiess (1929), p. 52

42 *Euler-Goldbach*

43 Fuss, II, p. 409f.; Mikhailov (1957), p. 24–26

44 Euler's autobiography, here p. 4. − About currency issues, cf. Fellmann (1992), p. 219–222

45 Mikhailov (1959), p. 275–278

46 The correspondence between Wolff and J. Bernoulli is worked on by F. Nagel in Basel and is to be edited as part of the Bernoulli-edition.

47 Mikhailov (1959), p. 276, Note 12

48 *Commentarii Academiae Imperialis scientiarum Petropolitanae*, I–XIV (1726–1746) 1728–1751

49 Kopelevich (1973), p. 121–133

50 About the complicated and tragigruesome history of Russia in the post Peter period, cf. for example Gitermann, Vol. 2, p. 143–295; v. Rimscha, p. 322–399

51 Gitermann, Vol. 2, p. 152

52 Lomonosov (1948; 1957), in the Bibliography under Vavilov

53 Kopelevich (1974), p. 176–229

54 Kopelevich (1977), Spiess (1929), p. 52f.

55 Bernoulli-Sutter (1972), p. 62f.

56 Cf. DSB VI; NDB 8. Only since recently has there been in print a complete bibliography of J. Hermann; see Nagel (1991).

57 *Euler's correspondence*, Part III, p. 66, Note 5

58 After Euler's departure to Berlin (1741), there ensued a correspondence between Euler and Krafft, which, in all respects, is very interesting and continued until 1753. It is published − for the most part only in the form of extracts or summaries − in *Euler's correspondence*, Part III, p. 134–176.

59 Gmelin (1747); Posselt (1990)

60 Cf. Nevskaya (1983), p. 363–371, especially the literature indicated there on p. 370 in the Notes 6 and 7

61 Cf. *Euler's correspondence*, Part I, p. 5

62 For a biography of G. F. Müller, cf. Büsching (1785), p. 1–160

63 *Euler's correspondence*, Part I

64 Goldbach was second godfather to Euler's first-born son Johann Albrecht.

65 An excellent biography exists in Yushkevich-Kopelevich (1983). It appeared in German translation by A. and W. Purkert with Birkhäuser, Basel 1994. Short expositions in *Euler-Goldbach*, p. 1–16; DSB V, article by M. S. Mahoney

66 In a special formulation, this conjecture of Goldbach, stated in the letter to Euler of June 7, 1742, says: Every even natural number not equal to 2 can be represented as sum of two prime numbers (Example: $8 = 5 + 3$; $52 = 47 + 5$). This statement, to this day, could not be proved in full generality, but it continues to have a stimulating influence on mathematical research into our days.

67 *Euler-Goldbach*

68 Spiess (1929), p. 63f.

69 Winter (1958)

70 E. 5: O.I,27

71 E. 6: O.II,6

72 E. 7: O.II,31

[73] D. Bernoulli (1738). For the early development of hydrodynamics and hydraulics under Daniel Bernoulli and Euler, cf. Mikhailov (1983)

[74] Euler's letter to Daniel Bernoulli of May 25, 1734 (R 98), published in *Bibl. Math.* (3), 7, p. 139

[75] *ibid.*

[76] Spiess (1929), p. 73

[77] *Registres*, p. 11

[78] Very recently, however, we discovered in the Euler-Archive of Basel among the diary entries of J. A. Euler the copy of a folio, which — next to geometrical sketches — shows a doodle of two figures, which could depict the parents of Euler (cf. the figure on p. 118)

[79] D. Bernoulli to Euler, March 29, 1738 (R 118). Possibly, the artist was the above-mentioned painter from Holstein, Georg Brucker, art master from 1735 to 1737 at the "Gymnasium" of the Petersburg Academy. The idea that the father-in-law Gsell has made the portraits cannot indeed be dismissed entirely, for the shipment of portraits contained also one of Tsarina Anna, as Bernoulli reports.

[80] About the pathology and pathogenesis of Euler's eye diseases, which eventually led to blindness, one may consult the currently authoritative account of R. Bernoulli (1983).

[81] D. Bernoulli to Euler, March 4, 1735 (R 100)

[82] With the exception of a single letter of Leonhard Euler, the entire correspondence with his parents, unfortunately, is lost.

[83] D. Bernoulli to Euler, November 8, 1738 (R 123)

[84] It was a matter of calculating atronomical tables to determine the meridian equation from two observed altitudes of the sun, in degrees, and under observation intervals of 1 to 18 hours for the pole altitude of St. Petersburg.

[85] Fuss (1783)

[86] *Euler-Goldbach*, p. 81

[87] Eneström-numbers 4–57

[88] In these sections, anticipations to later works of Euler are inevitable.

[89] Mikhailov (1985), p. 64–82

[90] Newton (1687)

[91] Hermann (1716)

[92] *Mechanica sive motus scientia analytice exposita*. Petersburg 1736 (E. 15,16; O.II,1,2). German transl. by J. Ph. Wolfers, Greifswald 1848–1850

[93] About the subdivision of mechanics, I refer to Hamel (1912), p. 9f. and also to the modern work Szabó (1975).

[94] *Theoria motus corporum solidorum seu rigidorum*...Rostock and Greifswald 1765 (E. 289: O.II,3)

[95] Maclaurin (1742)

[96] Lagrange (1788); Kovalevskaya (1889), p. 177–232

97   The most important and deep works about Euler's hydrodynamics are without doubt Truesdell (1954; 1956), which appeared as introductions to the Euler-volumes O.II,12,13. Cf. further Szabó (1987); Fellmann (1983b), especially p. 1122–1129; Truesdell (1968), especially p. 219–233; BV Bouligand (1960); BV Frankl (1950); BV Mikhailov (1960); BV Tyulina (1957); BV Truesdell (1957, 1960–1962, 1967, 1981); BV Szabó (1972)

98   *Scientia navalis seu tractatus de construendis ac dirigendis navibus* ... Petersburg 1749 (E. 110,111: O.II,18,19)

99   Habicht (1983a), p. 243

100   Cf. Habicht (1974) and (1978)

101   Johann Andreas Segner (1704–1777) was an important physicist, mathematician, and physician, cf. DSB XII, article by A. P. Yushkevich and A. T. Grigor'yan; Kaiser (1977). The 162 extant letters of Segner to Euler are of high scientific-historical interest, but still are not yet published. Cf. O.IVA,1 (R 2417–2575)

102   Ackeret (1944)

103   This subchapter is a slightly edited version of the corresponding sections in EGB83, p. 73–80. Participating on this were Beatrice Bosshart and Eugen Dombois. Cf. Fellmann (1983a)

104   E. 33: O.III,1. The fact that the book was written already in 1731 we know from a letter (R 199) of Euler to Johann Bernoulli. – A French translation of the *Tentamen* appeared in 1839 under the title *Essai d'une nouvelle théorie de la musique* in: *Œuvres complètes d'Euler*, 5, 1839, p. I–VII, p. 1–215.

105   E. 314,315,457: O.III,1. These works were submitted in the years 1760, 1764, and 1773.

106   Examples can be found in the secondary literature, for example in Helmholtz (1913), p. 377; E. Bernoulli in the preface to O.III,1; BV (here Note 124)

107   I do not know of any textbook in physics, in which the invention and use of the "equal temperament" with the constant half-tone interval $i = \sqrt[12]{2}$ is not attributed to Werckmeister (1691). This attribution is absolutely false and does not at all do justice to the able musician and sensitive theoretician Werckmeister (cf. Note 115, especially the work BV Kelletat)

108   BV Busch

109   Op. cit. BV Vogel

110   R 210. Based on the translation from Latin by EAF and Beatrice Bosshart

111   What is meant is the *Tentamen*

112   One has $96 = 2^5 \cdot 3$; $108 = 2^2 \cdot 3^3$; $120 = 2^3 \cdot 3 \cdot 5$; $128 = 1 \cdot 2^7$; $135 = 3^3 \cdot 5$; $144 = 2^4 \cdot 3^2$; $160 = 2^5 \cdot 5$; $180 = 2^2 \cdot 3^2 \cdot 5$; $192 = 2^6 \cdot 3$. In this way indeed – disregarding the tone $F^\#$ – there result the pure major intervals $24 : 27 : 30 : 32 : 36 : 40 : 45 : 48$. The half-tone step $G : F^\#$ is therefore equal to the "natural" value $F : E = 16 : 15 = 1.06667$, the half-tone interval $F^\# : F$, however, becomes 1.0546875, slightly smaller than the half-tone step $\sqrt[12]{2} = 1.05946$ in the equal temperament. – With regard to Euler's theory of music, compare E. Bernoulli's introduction to O.III,1 and also BV Busch.

113   According to the system of Mattheson, the interval $B^\flat : A$ with value $27 : 25 = 1.08$ (133 Cents) is a diatonic half-tone, according to Euler's, however, $B^\flat : A = F^\# : F$ with value $135 : 128 = 1.0546875$ (92 Cents) is a chromatic half-tone.

114   R 211. Translated from Latin by EAF and Beatrice Bosshart

[115] With the *progressio geometrica* is meant the equal temperament. In it, as is known, the frequencies of successive (half-) tones form a geometric sequence. Cf. EGB, p. 92, Note 135 and 141

[116] R 126

[117] J. Mattheson (1731), p. 139

[118] R 126

[119] It is probably Emanuel Pfaff (1701–1755), the head of the Collegium Musicum in Basel at that time.

[120] R 140

[121] This treatment is due to Beatrice Bosshart.

[122] Cf. Kelletat (BV), p. 321, and also the tables in the *Appendix* therein. However, in the Euler-table, the entry in row 4 and column 3 must be corrected: instead of $F\,8:9$, there should be $F^{\#}\,225:256$.

[123] BV Fellmann 1975

[124] Further literature: BV Bailey, Busch, Jacobi, E. R. Kalla-Heger, Kelletat, Mitzler, Scriba, Smith, Thiele, Ucello, Vogel, Winkel. In the (largely unpublished) manuscripts of Euler, which are deposited in the Archive of St. Petersburg (F. 136, op. 1), one can find music-theoretical notes on 51 pages of four of the twelve "note books", which for the most part have not appeared in Euler's printed works, but are of great importance. Eberhard Knobloch (Berlin) has subjected these manuscripts to a thorough analysis and reported his findings in an excellent paper: E. Knobloch (1987).

[125] Cf. Gitermann, Vol. 2, p. 160f.; v. Rimscha, p. 329f.

[126] *Registres*, p. 14

[127] *Euler's correspondence*, Part I, p. 226

[128] Condorcet (1783) 1786, p. 306

[129] Vladimir Grigorevich Orlov (1743–1831) was director of the Petersburg Academy during the years 1766–1774.

[130] The drama of those days has been recorded impressively by Euler himself, in chronological order, in his *Journal*, first published by W. Stieda (Stieda [1931], p. 8–10). This *Journal, betreffend die von Ihro Königl. Majestät von Preussen an mich ergangene Allergnädigste Vocation* [Journal, relating to the most gracious call come to me by Your Royal Majesty of Prussia] is printed in O.IVA,6, p. 298–301.

[131] The participation of Euler's brother Johann Heinrich (cf. Note 10) in the travelling party to our knowledge is not documented, yet very probable (cf. the work *Rasvitiye*, p. 472, cited in the Note 5).

[132] Initially, Frederick II offered a salary of 1,000–1,200 Thl., but Euler considered this too low compared to the Russian salary of 1,200 Rbl., whereupon it eventually was raised by 400 Thl. Cf. Harnack, I, p. 257, footnote 1

[133] Euler himself gave a rather detailed travel report in his letter of August 1, 1741 to Goldbach. In this letter, Euler also complains about Schumacher, who, in connection with the sale of Euler's house in Petersburg, left to him in trust, was to have tried in a dishonest manner to enrich himself by 100 Rbl. (*Euler-Goldbach*, p. 82–84; *Euler's correspondence*, Part II, p. 54f.).

[134] On this, D. Bernoulli was only too right. The new Academy — after a long prehistory — was officially opened only at the beginning of 1746. Of the three Bernoulli, none ended up accepting the call.

135  The First Silesian War ended with the peace treaty of Wrocław of June 11, 1742, the Second with the peace treaty of Dresden of December 25, 1745.

136  Translation from the Latin by EAF, cf. Fellmann (1992), p. 209f. Euler has not returned to Switzerland for the rest of his life.

137  Cf. O.IVA,6, p. 297

138  For all historical matters relating to the Academy, consult Harnack's standard work. An excellent brief account can be found in Dunken (1958). The avoidance of the notion of an "Academy" was Leibniz's well-founded intention.

139  The father of Frederick II, he reigned during 1713–1740.

140  Dunken (1958), p. 13

141  Cf. Harnack, I, p. 247–316. Euler's part in this merger is considerable.

142  *Miscellanea Berolinensia ad incrementum scientiarum ex scriptis Societatis Regiae scientiarum exhibtis edita.* I–VII, Berol. 1710–1744. About the content of these volumes, cf. *Harnack,* I, p. 235f.

143  *Euler-Goldbach,* p. 131

144  *Registres,* p. 23. Here, also the original titles of Euler's seven memoirs are listed.

145  E. 58: O.II,28. This is the comet which on February 8, 1742, passed through the perihelion and was observed in Paris from March 5 till May 6, 1742, by Cassini, Maraldi and Lacaille; its period is 164 years. (I thank Professor G. A. Tammann, Basel, for this kind information; EAF). Euler, however, relied on observational data which were furnished to him by Delisle in Petersburg.

146  E. 59: O.I,17

147  E. 60: O.I,17

148  E. 61: O.I,14

149  E. 62: O.I,22

150  E. 83: O.I,27

151  E. 284: O.I,22

152  *Registres,* p. 23f.; W. Knobloch (1984), p. 345. The memoir E. 83 then appeared in the first volume of the (new) academy letters *Mém. Berlin* (1745) 1746 as a translation into French, and E. 284 in NCP 9 (1762/1763) 1764 in the original Latin version.

153  *Euler-Goldbach,* p. 129f.

154  Born on November 15, 1741

155  Born on May 1, 1743

156  *Euler-Goldbach,* p. 186

157  Of course, one has to distinguish between the time of writing and the years of print. Cf. Eneström, p. 223–270, especially p. 227f. for the time span considered here.

158  *Commentatio de matheseos sublimioris utilitate* (E. 790: O.III,2). Here also a French version by E. Lévy (1853) and a German translation by J. J. Burckhardt (1942) is printed.

159  E. 27: O.I,25. It was presented to the Petersburg Academy on October 27, 1732 and printed 1738 in the CP 6. For a critical assessment, cf. C. Carathéodory, O.I,24, p. XXVIIf.

[160] L. Euler: *Methodus inveniendi lineas curvas maximi minimive proprietate gaudentes, sive solutio problematis isoperimetrici latissimo sensu accepti.* Lausanne, Geneva 1744 (E. 65: O.I,24). German translation (in extracts) by P. Stäckel: *Methode Curven zu finden, denen eine Eigenschaft im höchsten oder geringsten Grade zukommt, oder Lösung des isoperimetrischen Problems, wenn es im weitesten Sinne des Wortes aufgefaßt wird.* Leipzig 1894, OK nr. 46, p. 21–132. Notes of Stäckel, p. 133–143.

[161] For the history of the calculus of variations, consult, for example, Giesel (1857); Carathéodory (1945); BV Kneser (1907); Goldstine (1980); Hildebrandt (1984); Fraser (1992).

[162] Cf. Correspondence Euler-Lagrange, O.IVA,5, "Introduction" and p. 359f.

[163] L. Euler: *Elementa calculi variationum* (E. 296: O.I,25). This memoir was presented to the Petersburg Academy on December 1, 1760, appeared in print, however, only in 1766 in the NCP 10. In its title, the term "calculus of variations" appears for the first time.

[164] Cf. Biermann (1983), p. 489–500

[165] About the history of (outer) ballistics, cf. Szabó (1987), p. 199–224, about the subject Robins-Euler, especially p. 211–220.

[166] Cf. Note 85

[167] O.I,1, p. LXV

[168] Niklaus Fuss (1755–1826), a highly-talented young man, born in Basel of lowly origin, was "discovered" by Daniel Bernoulli when only seventeen years old and recommended to Euler as assistant to Petersburg, where, since 1773, he quickly made a career for himself as a close collaborator of Euler. Still in the last year of Euler's life, Fuss became an academician and professor of mathematics, married in 1784 the oldest daughter of Euler's son Johann Albrecht and acted since 1800 in the latter's succession as conference secretary of the Petersburg Academy. His son, Paul Heinrich Fuss (1798–1855), succeeded him in this position; together with C. G. J. Jacobi, he became engaged in the edition of Euler's works and, among other things, prepared (1843) the two-volume edition of letters (Fuss I,II). Cf. DSB V, article by A. I. Volodarskiy; *Lexikon*, p. 160f.

[169] Cf. Note 92

[170] Robins (1739), p. 1–29. Benjamin Robins (1707 1751) gained a name for himself principally as the inventor of the "ballistic pendulum". His foundation of "inner ballistics", which deals with the explosive force of gunpowder, and which in turn determines the initial velocity of the projectile, was quite new and original at the time. DSB XI, article by J. Morton Briggs, Jr.

[171] Cited after Szabó (1987), p. 210f.

[172] Robins (1742)

[173] Only J. H. Lambert succeeded in obtaining the explicit formula of the ballistic curve as an infinite series, and published it in *Mém. Berlin*, 21, (1765) 1767. Cf. Szabó (1987), p. 220–224.

[174] E. 77: O.II,14. The complete title takes up seven printed lines.

[175] A Russian translation appeared only in 1961.

[176] What is meant is the work indicated in Note 170.

[177] JB 135

[178] R 226. Translation from the Latin by EAF and Beatrice Bosshart.

[179] *Theoria motuum planetarum et cometarum.* Berlin 1744 (E. 66: O.II,28). About Euler's achievements in the area of celestial mechanics, see the contribution by Volk (1983).

[180] E. 67 and E. 68: O,II,31

[181] *Nova theoria lucis et colorum,* (E. 88: O.III,5)

[182] Speiser (1983), p. 216. Cf. especially D. Speiser's introduction to the volume O.III,5.

[183] *Introductio in analysin infinitorum.* Lausanne 1748 (2 Vols.). (E. 101,102: O.I,8,9). For additional editions, in particular translations, cf. O.I,8, p. XI. In German: Michelsen 1788; Maser (Vol. 1) 1885, 1983

[184] *Institutiones calculi differentialis...* Petersburg 1755 (2 Parts). (E. 212: O.I,10). German translation: Michelsen, Berlin 1790–1793

[185] *Institutionum calculi integralis...* Petersburg 1768–1794 (4 Vols.). (E. 342, 366, 385, 660: O.I, 11–13).

[186] Cf. the beautiful and easily accessible anthology Yushkevich (1983) in OK nr. 261; further Gel'fond (1983), p. 99–110, and also the extensive historical treatment Yushkevich (1976/77). An excellent survey on the *Introductio* is given in Cantor (1901), p. 699–721.

[187] *Euler-Goldbach,* p. 199. The orthography, there, is somewhat modernized.

[188] A. Krazer and F. Rudio in O.I,8, p. VIIIf.

[189] O.I,9, p. XIX

[190] Cf. Breidert (1983)

[191] *Lettres à une Princesse d'Allemagne sur divers sujets de physique et de philosophie.* Petersburg 1768–1772 (3 Vols.). (E. 343,344,417: O.III,11,12). Reprint of the first German edition, translated by Euler himself: Braunschweig 1986

[192] Sophie Friederike Charlotte Leopoldine von Brandenburg-Schwedt (1745–1808) (her given names are handed down differently) was the daughter of the Margrave Friedrich Heinrich von Brandenburg-Schwedt and a second cousin of King Frederick II. Since 1765, she was Abbess of the Convent of Herford.

[193] The Eneström-catalogue contains 111 different editions of the *Lettres*. Very easily accessible is the new French original edition, based on O.III 11,12, in a single volume, recently edited and admirably prefaced by S. D. Chatterji: Presses polytechniques et universitaires romandes, Lausanne 2003; the most recent English translation, in two volumes, prefaced by Andrew Pyle, appeared as reprint with Thoemmes Press, Bristol/Washington 1997 and goes back — as does the reprint of Arno Press, New York 1975 — to the first English translations of H. Hunter (1795, 1802) and D. Brewster (1823, 1833). The first German edition — probably in a translation by Euler himself — appeared in Leipzig 1769, of which Vieweg, Braunschweig 1986 published a reprint in one volume.

[194] With regard to Berkeley, there exists an excellent work biography in the German language: Breidert (1989).

[195] D'Alembert to Lagrange. Translated by EAF based on a French-Latin quotation in Spiess (1929), p. 119f.

[196] D. Bernoulli to Euler, April 29, 1747 (R 170)

[197] E. 292: O.III,12

[198] The facts are sketched briefly and expertly in Breidert (1983), p. 448f.

[199] *Réflexions sur l'espace et le temps* (E. 149: O.III,2)

[200] Speiser (1934)

201  Riehl (1908); Timerding (1919)
202  Alexander Gottlieb Baumgarten (1714–1762), a pupil of Christian Wolff and pro-
     fessor in Halle and Frankfurt (Oder), introduced in Germany aesthetics – in the
     sense of a theory of perception – as a discipline of philosophy, and is considered an
     important link between Wolff and Kant in the development of the philosophy of
     perception.
203  The text of the prize question can be found in *Harnack*, II, p. 305. The prize was won
     by J. H. G. Justi (1720–1771), a lawyer in Sangerhausen, with a "rather weak paper"
     (Spiess [1929], p. 119), which – rightly so – was perceived by the "Wolffians" as a
     scandal.
204  Letter of July 4, 1747, to Goldbach (*Euler-Goldbach*, p. 275f.)
205  Fellmann (1983a), p. 73; Fellmann (1975), p. 519
206  This personality, to this day, could not be identified.
207  *Euler-Goldbach*, p. 200
208  François André Danican Philidor (1726–1795) originally was a musician and, since
     1760, was considered the principal representative of comic opera in France. Already
     at the age of 20, he was one of the strongest chess players of his time. Philidor's
     myth, however, sustained damage by Van der Linde (1874), Vol. I, p 383–412.
209  *Euler-Goldbach*, p. 334
210  Philidor (1749). In this book, Philidor laid down his fundamental (not always sound)
     knowledge about chess strategy.
211  *Euler-Goldbach*, p. 336f.
212  "Opposite squares" are to be understood along the main diagonals, thus 33–1, 59–
     27, 45–13, 54–22 etc., but also squares oriented central-symmetrically, such as 58–26
     etc., are subsumed under this term.
213  *Euler-Goldbach*, p. 394. There, the word *arealis* is to be replaced by *areolis*.
214  Cf. Ahrens (1921), p. 319f.; Van der Linde (1874), Vol. 2, p. 101
215  E. 309: O.I,7
216  E. 530,795: O.I,7
217  Carl Friedrich von Jaenisch (1813–1872), professor of mechanics in St. Petersburg,
     who was the most important chess theoretician in the 19th century and the founder
     of the "Russian school of chess". He authored a three-volume work, the second
     volume of which is dedicated to the knight's move. Cf. Jaenisch (1862)
218  *Harnack*, I, p. 249
219  *Harnack*, I, p. 255
220  *Harnack*, I, p. 256
221  *Harnack*, I, p. 257
222  The reader can find the relevant literature in the Bibliography in O.IVA,6, p. 404–
     418, under the names Maupertuis, Angliviel de la Beaumelle, Brunet, Le Sueur.
223  Even more successful as popularizer was probably only Voltaire, who, from 1726
     on, stayed there for three years as *réfugié*. Cf. the excellent standard biography by
     Bestermann (1971), especially Ch. 11
224  A testimony of this friendship is the (not yet edited) correspondence between Mau-
     pertuis and Johann II Bernoulli, comprising 169 letters of Maupertuis and nine of
     Bernoulli (University Library Basel, Ms L Ia 662,676,708 and L Ib 90). This correspon-
     dence was utilized by P. Costabel in O.IVA,6, *passim*.

225  Cf. DSB V, p. 434–436, article by Seymor L. Chapin

226  Cf. DSB XV, p. 269–273, article by Yves Laissus

227  Maupertuis (1738) and König (1741)

228  Spiess (1929), p. 86f.

229  *Harnack*, I, p. 258

230  Cf. Spiess (1929), p. 82

231  For a description of the war activities in the large, cf. Mittenzwei (1984), p. 55–80; for the specific, cf. Groehler (1889), p. 22–63

232  Cf. *Harnack*, I, p. 293

233  1692–1769, since 1754 honorary member of the Berlin Academy (cf. NDB). Co-stabel's statement in O.IVA,6, p. 5 that Maupertuis's father-in-law is identical with Kaspar Wilhelm von Borcke, one of the four curators of the Berlin Academy, is false. For the clarification of the facts, I thank Dr. Fritz Nagel of the Bernoulli-edition in Basel and Martin Mattmüller of the Euler-Committee of the Swiss Academy of Sciences, as also for the establishment of the exact time in which Maupertuis was present in Basel.

234  *Harnack*, I, p. 299

235  *Registres*, p. 41

236  This correspondence, through a critical analysis of the texts, was published by P. Costabel in O.IVA,6. It comprises 137 pieces of letters and, unfortunately, for the most part only one way of the correspondence has survived. Except for seven letters, all of Maupertuis's letters to Euler are missing, since the latter, upon Mau-pertuis's explicit wish, has destroyed them – along with his own drafts and copies. Cf. O.IVA,6, p. 5, Note 4

237  Maupertuis to Johann II Bernoulli on January 19, 1745 (cf. Note 224)

238  Cf. *Registres*, p. 95f.

239  As to the character of Euler, I refer to the relevant comments in the Prologue.

240  The development and substance of this controversy has already been described many times. The newest (and most detailed) account of this theme is in Pulte (1989). Consult there the bibliography, especially under the authors Costabel, Fleckenstein, Graf, Kneser, Szabó.

241  *Registres*, p. 39. Cf. *Euler-Goldbach*, p. 199

242  Letter of November 24, 1750 (translation EAF), cf. Note 224

243  As to the question with what right Frederick II can be provided with the ornate epithet "the Great", Ingrid Mittenzwei has dedicated a whole chapter in her excellent Frederick-biography (see therein p. 80f.). Cf. also the very critical, well-substantiated account by R. Augstein (1986).

244  *Euler's correspondence*, Part I, p. 45

245  This correspondence is published in O.IVA,6, though somewhat thinly annotated, by E. Winter. It comprises 87 items of letters in the German and French language, of which 28 by Euler.

246  Allusion to Euler's *Gunnery* (cf. p. 65)

247  *Euler's correspondence*, Part I, p. 3

248  Mittenzwei, p. 84

249  Euler's correspondence with Frederick II., O.IVA,6

250  *Sur la perfection des verres objectifs des lunettes.* E. 118: O.III,6

[251] What is meant are the reflector telescopes of Newton and Gregory.

[252] *Euler-Goldbach*, p. 304

[253] Goethe (1949), p. 646

[254] *Nova theoria lucis et colorum*. Berlin 1746, E. 88: O.III,5. In this paper, Euler explored a possible isomorphism between light and sound waves. This is why Goethe may have been fascinated especially by Euler's ideas about the "resonance of colors".

[255] Cf. DSB XIV, p. 562–573, article by E. W. Morse

[256] Most recent investigations show, however, that Newton had also a positive attitude toward achromatism.

[257] The chromatic aberation of the human eye was rigorously established only in 1817 by Joseph Fraunhofer. Cf. DSB V, p. 142–144, article by R. V. Jenkins

[258] E. 460: O.III,8, p. 202

[259] A detailed account of discoveries and of Euler's position in the history of optics can be found in Fellmann (1983c).

[260] E. 367,386,404: O.III,3,4

[261] E. 844,844a: O.III,9. The mathematical substance of this work — rendered by W. Habicht in a concise modern language — can be found in Habicht (1983b).

[262] *Euler's correspondence*, Part II, p. 63

[263] *Euler's corrrespondence*, Part II, p. 62, Note 2

[264] *Euler's correspondence*, Part II, p. 63 and Plates I and II with facsimile of Euler's (Russian) handwriting

[265] The two-volume *Theory of ships*; it was then (1749) printed in St. Petersburg after all.

[266] Letter of August 13/24, 1743, *Euler-Goldbach*, p. 179

[267] *Euler-Goldbach*, p. 195

[268] *Euler-Goldbach*, p. 200

[269] *Euler's correspondence*, Part III, p. 275. Euler's correspondence, comprising 57 items, with the theologian Caspar Wettstein in Basel — with the sole exception of a letter of Wettstein to Euler (R 2755) — has survived only as a one-way correspondence. It was conducted entirely in French and has been edited almost completely. Cf. the summaries in O.IVA,1, R 2748–2804.

[270] Cf. Fellmann (1992), p. 219f.

[271] *Euler-Goldbach*, p. 311

[272] Letter of July 27/August 7, 1753 to Schumacher, *Euler's correspondence*, Part II, p. 318

[273] Letter of October 7/18, 1760 to Müller, *Euler's correspondence*, Part II, p. 161f. — Euler was later indemnified.

[274] About him, there are reports by Stäckel (1910) and Stieda (1932), as also by Jaquel (1983).

[275] Letter of May 6/17, 1754 to Goldbach, *Euler-Goldbach*, p. 377

[276] Letter of April 15/26, 1754 to Euler, *Euler-Goldbach*, p. 376. Goldbach possibly alludes to a facetious-ironic utterance — to this day not documented — of Jakob Hermann, which the latter has made, perhaps in connection with Daniel Bernoulli's custom to add to his name the epithet "Johannis filius". This occurred, for the last time, on the title page of the *Hydrodynamica* of 1738.

[277] E. A 7. Cf. *Euler-Goldbach*, p. 397

[278] Cf. DSB, XII: Kaiser (1977)

[279] Cf. O.IVA,1, R 2417–2575

Of this "mammoth correspondence", only the 159 letters of Segner have survived: the latter, unfortunately, had directed, by will, to have his entire literary remains burned posthumously.

[280] O.IVA,6, p. 384. The marriage took place only after the transfer of the Euler family to St. Petersburg (1766). Cf. K. Euler (1955), p. 273

[281] *Euler's correspondence*, Part II, p. 433

[282] Stieda (1931); *Registres*, Introduction; *Euler's correspondence*, Parts I–III; Yushkevich (1982); Biermann (1983); *Harnack*, I

[283] About d'Alembert, cf. DSB I, article by J. Morton-Briggs

[284] *Euler's correspondence*, Part I, p. 71, p. 213, p. 230, p. 232: *Registres*, p. 70f.

[285] *Euler-Goldbach*, p. 400

[286] O.IVA,5, letters nr. 37a and 38

[287] *Euler's correspondence*, Part I, p. 225

[288] Cf. Hartweg (1979), nr. 4, p. 14f.; nr. 5, p. 17f.

[289] *Registres*, p. 90

[290] *Harnack*, I, p. 363f., *Registres*, p. 74–76; Biermann (1985), p. 93–95

[291] Cf. Biermann (1985). About Lambert, cf. *Colloque Lambert* 1977; Jaquel (1977); DSB VII, article by C. J. Scriba

[292] O.IVA,6, R 695–698

[293] Cited after *Registres*, p. 86f.

[294] Translation: "With reference to your letter of April 30, I am giving you permission to resign in order to go to Russia. Federic."

[295] About Lagrange, cf. DSB VII, article by J. Itard; O.IVA,5, Introduction, p. 34–63

[296] This correspondence — a fascination for every mathematician — is critically published in O.IVA,5.

[297] Frederick II (1789), p. 13f.

[298] *Pekarskiy*, I, p. 303

[299] Spiess (1929), p. 185f.

[300] Cf. Note 80

[301] Mikhailov (1959), p. 274

[302] *Harnack*, I, p. 473

[303] Details in Stieda (1931), p. 38–41

[304] O.IVA,6, p. 393–396

[305] E. 387: O.I,1, edited by H. Weber

[306] EGB 83, p. 84, Note 50

[307] Reich (1992), p. 148f. The last Reclam-edition was attended to by J. E. Hofmann 1959 in linguistically modernized form.

[308] Cf. the articles Volk (1983) and Nevskaya (1983)

[309] *Theoria motus lunae*. Berlin 1753, E. 187: O.II,23

[310] E. 112: O.II,25

[311] Cf. DSB IX, article by E. G. Forbes

[312] *Theoria motuum lunae*. Petersburg 1772, E. 418: O.II,22

[313] Cf. DSB VII, article by C. Eisele

[314] This chapter is an original work, slightly abridged (by me, EAF), of Professor
G. K. Mikhailov, which the latter, one of the best experts of Euler's manuscripts,
has kindly made available to me for publication. G. K. Mikhailov, for many years
a close collaborator of the Euler-edition series IV, is a member of the International
Editorial Committee.

[315] Stieda (1932), p. 26

[316] Dashkov (1857), p. 36

[317] Burya (1785)

[318] E. 579: O.II,16. Cf. also Ackeret (1945)

[319] Condorcet (1783) 1786, p. 309 in O.III,12

[320] Cf. Petrov (1958)

[321] Cf. Fellmann/Im Hof (1993)

[322] Diderot (1961), p. 421

[323] Schmieden/Laugwitz (1985); Laugwitz (1978; 1983; 1986) and the literature therein

# Chronological table

The years of publication of Euler's principal works are not indicated here (we refer to the Bibliography, Section 3). For dates of birth and death of Euler's children, consult the compilation on p. 105.

1707    Leonhard Euler is born on April 15 in Basel, the son of the reformed minister Paulus Euler and of Margaretha Brucker.

1708    Paulus Euler assumes the incumbency in Riehen near Basel.

1713    After private tuition by the father, attendance of the Latin school ("Gymnasium") in Basel.

1720    Begin of studies at the University of Basel

1722    *Prima Laurea* (lowest academic degree, corresponding roughly to today's Matura, which at that time had to be earned during the first semesters at the university). Freshman courses in mathematics from Johann I Bernoulli.

1723    Autumn: Promotion to Magister (conclusion of studies at the Philosophical Faculty). Enrollment at the Theological Faculty. *Privatissima* (privat lessons) with Johann I Bernoulli.

1724    Public speech on the systems of Descartes and Newton.

1725    Death of Tsar Peter I. His wife Catherine I ascends to the Russian throne. Foundation of the Petersburg Academy, to which Johann I Bernoulli's sons Daniel and Niklaus committed themselves as professors and travelled thereto, as also the mathematician Jakob Hermann from Basel.

1726    Euler's first mathematical work appears in print in Leipzig.

1727    Euler participates at a prize question of the Paris Academy and earns second prize (*accessit*) with his paper on the optimal position of a mast on a ship. He competes without success for the profesorship of physics in Basel and accepts a call to the Petersburg Academy, where he starts his career as an adjunct. A week before Euler's arrival Tsarina CatherineI dies.

1730    Jakob Hermann returns to Basel; Daniel Bernoulli takes over the latter's chair of mathematics at the Petersburg Academy. After Peter's II short reign, Anna Ivanovna becomes tsarina for the next ten years.

1731    Euler becomes professor of physics, succeeding Georg Bernhard Bülfinger, and at the same time is promoted to an ordinary member of the Petersburg Academy.

1733    Daniel Bernoulli returns to Basel, and Euler takes over the latter's professorship of mathematics in Petersburg. Jakob Hermann dies in Basel. Beginning of the great Kamchatka-expedition, which is to last ten years.

1734    Euler on January 7 (December 27, 1733, old style) marries Katharina Gsell. Birth of the first son Johann Albrecht on November 27.

1735    Euler shares (with Delisle) the direction of the Department of Geography at the Petersburg Academy and collaborates actively on the cartography of Russia.

1738    Euler looses his right eye as a result of a dangerous disease.

1740    Death of Tsarina Anna Ivanovna. In Prussia King Frederick Wilhelm I dies, and his son ascends to the throne as Frederick II. Maria Theresia becomes Queen of Austria and Hungary. Beginning of the First Silesian War, started by Frederick II, which lasts for two years.

1741    A coup d'état in Russia brings to power for the next twenty years Elisabeth Petrovna, the daughter of Peter I. In the summer, Euler accepts the invitation of Frederick II to come to Berlin in order to help there with the establishment of an Academy.

1744/45    The Second Silesian War delays the establishment of the Berlin Academy. The war lasts until 1745.

1745    Euler's father Paulus dies in Riehen (Basel). Maupertuis comes to Berlin.

1746    January: The Berlin Academy officially opens with Maupertuis as President and Euler as Director of the Mathematical Class. Euler becomes Fellow of the London Royal Society.

1748    Johann I Bernoulli dies in Basel.

1749    First personal encounter of Euler with Frederick II.

1750    Euler meets his mother in Frankfurt a. M., in order to take her with him to Berlin.

1754    Euler's son Johann Albrecht becomes member of the Berlin Academy.

1755    Leonhard Euler becomes a foreign member of the Paris Academy.

1756    Outbreak of the Seven-Year War.

1759    Maupertuis, since 1753 for all practical purposes substituted by Euler as president of the Berlin Academy, dies in Basel.

1762    Catherine II – following the death of Tsarina Elisabeth and the removal of her husband Peter III – ascends to the Russian throne. She invited Euler to return to St. Petersburg.

1763    End of the Seven-Year War with the peace treaty of Hubertusburg. D'Alembert, upon the invitation of Frederick II, stops in Berlin, declined, however, the presidency of the Berlin Academy that had been offered to him. Euler is already negotiating with the Petersburg Academy about his return.

1766    Falling out between Frederick II and Euler. In the summer, the latter returns with his family to Petersburg, where he is given a triumphant reception. Lagrange becomes Euler's successor at the Berlin Academy. More and more, Euler suffers from a decline of his visual faculty in his left eye due to a cataract.

1771    A cataract operation leads, after a short time, to the loss also of his left eye. Fire of Petersburg; Euler looses his house which, however, is replaced by the tsarina.

1773    Death of Euler's wife. Niklaus Fuss comes from Basel to Petersburg as Euler's assistant.

1773–75 Pugachov-revolt in Russia.

1776    Euler marries Salome Abigail Gsell, a sister-in-law of his first wife.

1782    Daniel Bernoulli dies in Basel on March 17.

1783    On September 18, Euler suffers a stroke and dies quickly and painlessly.

# Bibliography

## 1. Abbreviations used throughout

| | |
|---|---|
| AAN: | Archive of the Akademiya Nauk in St. Petersburg |
| AHES: | *Archive for History of Exact Sciences* |
| AP: | Acta Petropolitana |
| *Bibl. math.*: | *Bibliotheca mathematica* |
| *Euler's correspondence*: | Yushkevich-Winter, 3 Parts, 1959–1976 |
| BV: | Burckhardt-Verzeichnis in EGB 83 |
| CAP: | *Commentarii Academiae Petropolitanae* |
| DSB: | *Dictionary of Scientific Biography* (cf. Gillispie) |
| E: | Eneström-catalogue, with nr. following it |
| EGB 83: | *Euler-Gedenkband*. Basel 1983 |
| *Euler-Goldbach*: | Yushkevich-Winter 1965 |
| Fuss I,II: | *Correspondance 1843* |
| *Harnack*: | History of the Berlin Academy |
| HM: | *Historia Mathematica* |
| IMI: | *Istoriko-Matematicheskiye Issledovanniya*, 1948–2006, 45 vols. (Russian) |
| JB: | Johann Bernoulli, with work nr. following it according to JBO |
| JBO: | Johann Bernoulli: *Opera* |
| *Lexikon*: | *Lexikon bedeutender Mathematiker*. Thun, Frankfurt a. M. 1990 |
| *Mém. Berlin*: | *Histoire et mémoires de l'Académie de Berlin* |
| NCP: | *Novi Commentarii Academiae Petropolitanae* |
| NDB: | *Neue Deutsche Biographie* |
| O: | *Opera omnia Leonhardi Euleri* |
| OK: | Ostwalds Klassiker der exakten Wissenschaften |
| R: | Résumé-nr. from O.IVA,1 |
| *Registres*: | Winter 1957 |

## 2. Bibliographies

Eneström, Gustaf: Verzeichnis der Schriften Leonhard Eulers. In: *Jahresbericht der Deutschen Mathematiker-Vereinigung*. Ergänzungsband 4, 1. Lieferung 1910. 2. Lieferung 1913. Leipzig 1910–1913

159

Burckhardt, Johann Jakob: *Euleriana—Verzeichnis des Schrifttums über Leonhard Euler.* In: EGB 83, p. 511–552

Verdun, Andreas: *Bibliographia Euleriana.* Bern 1998

O.IVA,1: Register volume of Euler's correspondence

# 3. Works

*Leonhardi Euleri Opera Omnia.* Edited by the Euler-Committee of the Swiss Academy of Sciences (formerly Swiss Society of Natural Sciences). Leipzig and Berlin 1911f., Zurich, Basel 1982f. Series I: *Opera mathematica.* 32 vols. in 29 (all appeared); Series II: *Opera mechanica et astronomica.* 32 vols. in 31 (appeared except for Vols. 26, 27); Series III: *Opera physica, Miscellania,* 12 vols. (all appeared); Series IVA: *Commercium epistolicum* (Correspondence), 9 vols., appeared Vol. 1: Register volume with various listings, Basel 1975; Vol. 2: Briefwechsel Eulers mit Johann (I) und Niklaus (I) Bernoulli, Basel 1998; Vol. 5: Euler's correspondence with Clairaut, d'Alembert and Lagrange, Basel 1980; Vol. 6: Euler's correspondence with Maupertuis and Frederick II, Basel 1986; Series IVB: *Manuscripta, Adversaria* (manuscripts, note books and diaries), ca. 7 vols. (none appeared yet). Series I–III constitute practically an edition in final form, Series IV is a historio-critical edition. The letters in Latin are each accompanied by a translation into a modern language.

Euler's major works (in abbreviated titles, ordered chronologically according to years of print).

| | |
|---|---|
| 1736: | *Mechanica* (2 vols.) |
| 1738 }<br>1740 : | *Rechenkunst* (2 vols.) |
| 1739: | *Tentamen novae theoriae musicae* (Music theory) |
| 1744: | *Methodus inveniendi* (Calculus of variation) |
| 1745: | *Neue Grundsätze der Artillerie* (Ballistics) |
| 1747: | *Rettung der göttlichen Offenbarung gegen die Einwürfe der Freygeister* |
| 1748: | *Introductio in analysin infinitorum* (Introduction to analysis, 2 vols.) |
| 1749: | *Scientia navalis* (Theory of ships, 2 vols.) |
| 1753: | *Theoria motus lunae* (First lunar theory) |
| 1755: | *Institutiones calculi differentialis* (Differential calculus, 2 vols.) |
| 1762: | *Constructio lentium objectivarum* (Achromatic lenses) |
| 1765: | *Theoria motus corporum* (Second mechanics) |
| 1768: | *Lettres à une Princesse d'Allemagne* (Philosophical letters, 3 vols.) |
| 1768: | *Institutiones calculi integralis* (Integral calculus, 3 vols. through 1770) |

1769:   *Dioptrica* (Universal optics, 3 vols. through 1771)

1770:   *Vollständige Anleitung zur Algebra* (2 vols.)

1772:   *Theoria motuum lunae* (Second lunar theory)

1773:   *Théorie complette de la construction et de la manœuvre des vaisseaux*
        (Second theory of ships)

Postscript: The respective Eneström numbers and the distribution onto the individual volumes of the *Opera omnia* can be gathered from the Notes.

# 4. Edition of correspondences

Fuss, Paul Heinrich: *Correspondance mathématique et physique de quelques célèbres geomètres du XVIIIème siècle.* St. Petersburg 1843, 2 vols.

Eneström, Gustaf: *Der Briefwechsel zwischen Leonhard Euler und Johann I Bernoulli.* In: *Bibliotheca mathematica* (3)4, 1903; (3)5, 1904; (3)6, 1905 (incomplete)

–: *Der Briefwechsel zwischen Leonhard Euler und Daniel Bernoulli.* In: *Bibliotheca mathematica* (3)7, 1906–1907 (incomplete)

Smirnov, V. I. et al.: *Leonard Eyler—Pis'ma k uchonym.* Moscow, Leningrad (AN SSSR) 1963 (Selection. Original texts with Russian translation and commentaries)

Yushkevich, A. P., Winter, E.: *Die Berliner und die Petersburger Akademie der Wissenschaften im Briefwechsel Leonhard Eulers.* 3 vols. Berlin 1959–1976. Part I: 1959; Part II: 1961; Part III: 1976

–: *Leonhard Euler und Christian Goldbach. Briefwechsel 1729–1764.* Berlin 1965

O.IVA,2: *Leonhard Euler: Briefwechsel mit Johann (I) Bernoulli und Niklaus (I) Bernoulli.* Basel 1998

O.IVA,5: *Correspondance de Leonhard Euler avec A. C. Clairaut, J. d'Alembert et J. L. Lagrange.* Basel 1980

O.IVA,6: *Correspondance de Leonhard Euler avec P.-L. de Maupertuis et Frédéric II.* Basel 1986

Forbes, Eric G.: *The Euler-Mayer correspondence (1751–1755).* New York 1971

Postscript: For other letters, edited individually, and correspondences of Euler, consult the register volume O.IVA,1.

# 5. Memorial publications and omnibus volumes (selection)

Festschrift zur Feier des 200. Geburtstages von Leonhard Euler. Edited by the Berlin Mathematical Society. Leipzig-Berlin 1907 (Abhandlungen zur Geschichte der mathematischen Wissenschaften, Fasc. XXV. Preface by P. Schafheitlin, E. Jahnke, C. Färber)

Leonard Eyler. Sbornik statey v chest' 250-letiya so dnya rozhdeniya, predstavlennykh Akademiy nauk SSSR. M. A. Lavrentev, A. P. Yushkevich, A. T. Grigoryan (Eds.). Moscow 1958 (Russian, with German summaries)

Sammelband der zu Ehren des 250. Geburtstages Leonhard Eulers der Deutschen Akademie der Wissenschaften zu Berlin vorgelegten Abhandlungen. K. Schröder (Ed.). Berlin 1959

Leonhard Euler 1707–1783. Beiträge zu Leben und Werk. Gedenkband des Kantons Basel-Stadt. J. J. Burckhardt, E. A. Fellmann, W. Habicht (Eds.). Basel 1983

Zum Werk Leonhard Eulers. Vorträge des Euler-Kolloquiums im Mai 1983 in Berlin. Edited by E. Knobloch, I. S. Louhivaara, J. Winkler. Basel, Boston, Stuttgart 1984

Festakt und Wissenschaftliche Konferenz aus Anlaß des 200. Todestages von Leonhard Euler. Edited by W. Engel. (Abhandlungen der Akademie der Wissenschaften der DDR. Abteilung Mathematik–Naturwissenschaften–Technik. Jg. 1985, Nr. 1 N). Berlin 1985

Razvitiye idey Leonarda Eylera i sovremennaya nauka. N. N. Bogolyubov, G. K. Mikhailov, A. P. Yushkevich (Eds.). Moscow 1988 (Russian)

# 6. Biographies (without dictionary-articles)

Du Pasquier, Louis-Gustave: *Léonard Euler et ses amis*. Paris 1927

Fuss, Nicolaus: *Lobrede auf Herrn Leonhard Euler*. In: O.I,1. (This elogy was read by Fuss in the French language on October 23, 1783 to the Petersburg Academy and published there in the same year. The German translation—prepared by Fuss himself—appeared in Basel 1786.)

Fellmann, Emil A.: *Leonhard Euler*. In: Kindler-Enzyklopädie "Die Großen der Weltgeschichte", Vol. 6. Zurich 1975, p. 469–531

–: *Leonhard Euler. Ein Essay über Leben und Werk*. In: EGB 83, p. 13–98 [Fellmann (1983a)]

Yushkevich, Adolf P.: *Leonard Eyler*. Moscow 1982 (Seriya Matematika, Kibernetika 7 [1982])

Spiess, Otto: *Leonhard Euler*. Frauenfeld, Leipzig 1929

Wolf, Rudolf: *Biographien zur Kulturgeschichte der Schweiz*. Vierter Cyclus. Zurich 1862, p. 87–134

A listing of books containing larger sections on L. Euler can be found in EGB 83, p. 551ff.

# 7. Secondary literature

Ackeret, Jakob: *Untersuchung einer nach Eulerschen Vorschlägen (1754) gebauten Wasserturbine.* In: Schweizerische Bauzeitung 123 (1944), p. 9–15

–: *Leonhard Eulers letzte Arbeit.* In: Festschrift zum 60. Geburtstag von Prof. Dr. Andreas Speiser. Zurich 1945, p. 160–168

Ahrens, Wilhelm: *Mathematische Unterhaltungen und Spiele.* Vol. 1, 3d ed. Leipzig, Berlin 1921

Aiton, Eric J.: *The Vortex Theory of Planetary Motions.* London, New York 1972

Augstein, Rudolf: *Preußens Friedrich und die Deutschen.* Nördlingen 1986

Bernoulli, Daniel: *Hydrodynamica, sive de viribus et motibus fluidorum commentarii.* Straßburg 1738 (Translations: Russian 1959; German 1965; English 1968)

Bernoulli, Jacob: *Opera.* 2 vols. Geneva 1744; reprint: Brussels 1967

Bernoulli, Johann: *Opera omnia.* 4 vols. Lausanne and Geneva 1742; reprint: Hildesheim 1968 with preface by J. E. Hofmann. In the Notes to our volume, works of J. Bernoulli are sometimes provided with work numbers of this edition.

Bernoulli, René: *Leonhard Eulers Augenkrankheiten.* In: EGB 83, p. 471–487

Bernoulli-Sutter, René: *Die Familie Bernoulli.* Basel 1972

Besterman, Theodore: *Voltaire.* Munich 1971

Biermann, Kurt-Reinhard: *Aus der Vorgeschichte der Euler-Ausgabe 1783–1907.* In: EGB 83, p. 489–500

–: *Wurde Leonhard Euler durch J. H. Lambert aus Berlin vertrieben?* In: Abhandlungen der Akademie der Wissenschaften der DDR – Abteilung Mathematik, Naturwissenschaften, Technik, Jg. 1985, Nr. 1 N, Festakt. . . Leonhard Euler. Berlin 1985

Börner, Friedrich: *Nachrichten von den Lebensumständen berühmter Männer. Johann Jakob Ritter.* Vol. 2. Wolfenbüttel 1752

Breidert, Wolfgang: *Leonhard Euler und die Philosophie.* In: EGB 83, p. 447–457

–: *George Berkeley 1685–1753.* Basel, Boston, Berlin 1989

Buganov, Victor: *Peter der Große.* Cologne 1989

Burckhardt-Biedermann, Th.: *Geschichte des Gymnasiums zu Basel.* Basel 1889

Burya, Abel: *Observations d'un voyageur sur la Russie, la Finlande, la Livonie, la Curlande et la Prusse.* Berlin 1785 and 1787

Büsching, A. F.: *Beyträge zur Lebensgeschichte denkwürdiger Personen. . . ,* Part III. Halle 1785

Cantor, Moritz: *Vorlesung über Geschichte der Mathematik.* Vol. 3, 2d ed. Leipzig 1901

Carathéodory, Constantin: *Basel und der Beginn der Variationsrechnung.* In: Festschrift zum 60. Geburtstag von Prof. Dr. Andreas Speiser. Zurich 1945, p. 1–18. Same in: Constantin Carathéodory: *Gesammelte Mathematische Schriften,* Vol. II. Munich 1955

*Colloque international et interdisciplinaire Jean-Henri Lambert.* Mulhouse 26–30 septembre 1977. Paris 1979

*Commentarii Academiae imperialis scientiarum Petropolitanae.* Vols. I–XIV. Petersburg (1726–1746) 1728–1751

Condorcet, Antoine Caritat Marquis de: *Éloge de M. Euler.* In: Histoire de l'Académie Royale des sciences. Paris (1783) 1786, p. 37–68; reprinted in O.III,12, p. 287–310

Dashkov: *Memoiren der Fürstin Dashkov. Zur Geschichte der Kaiserin Katharina II.* Translated and prefaced by A. Herzen. Second Part. Hamburg 1857

Diderot, Denis: *Gedanken zur Interpretation der Natur (1754).* In: Philosophische Schriften, Vol. 1, edited by Th. Lücke. Berlin 1961

Donnert, Erich: *Rußland im Zeitalter der Aufklärung.* Leipzig 1983

Dunken, Gerhard: *Die Deutsche Akademie der Wissenschaften zu Berlin in Vergangenheit und Gegenwart.* Berlin 1958

Euler, Karl: *Das Geschlecht der Euler-Schölpi.* Gießen 1955

Fellmann, Emil A.: *Leonhard Euler 1707–1783. Schlaglichter auf sein Leben und Werk.* In: Helvetica Physica Acta, Vol. 56 (1983), p. 1099–1131 [1983b]

–: *Leonhard Eulers Stellung in der Geschichte der Optik.* In: EGB 83, p. 303–329 [1983c]

–: *Über einige mathematische Sujets im Briefwechsel Leonhard Eulers mit Johann Bernoulli.* In: Verhand. Naturf. Ges. Basel 95 (1985). Basel 1985, p. 139–160

–: *Non-Mathematica im Briefwechsel Leonhard Eulers mit Johann Bernoulli.* In: Amphora. Festschrift für Hans Wussing zu seinem 65. Geburtstag. Basel, Boston, Berlin 1992, p. 189–228

–, Im Hof, Hans-Christoph: *Die Euler-Ausgabe. Ein Bericht zu ihrer Geschichte und ihrem aktuellen Stand.* In: Jahrbuch Überblicke Mathematik 1993. Braunschweig, Wiesbaden 1993, p. 185–198

Finster, Reinhard, van den Heuvel, Gerd: *Gottfried Wilhelm Leibniz.* Reinbek bei Hamburg 1990

Fraser, Craig C.: *Isoperimetric Problems in the Variational Calculus of Euler and Lagrange.* In: HM 19 (1992), No. 1, p. 4–23

*Friedrichs des Zweiten Königs von Preußen hinterlassene Werke.* Translated from French [anonymous], Vol. 11. New improved and enlarged edition, Berlin 1789

Gelfond, Aleksander O.: *Über einige charakteristische Züge in den Ideen L. Eulers auf dem Gebiet der mathematischen Analysis und seiner "Einführung in die Analysis des Unendlichen".* In: EGB 83, p. 99–110

Giesel, Carl Franz: *Geschichte der Variationsrechnung.* Torgau 1857

Gillispie, Charles Coulston (Ed.): *Dictionary of Scientific Biography.* 16 vols. New York 1970–1980; Vols. 17–18: F. L. Holmes (Ed.), New York 1990

Gitermann, Valentin: *Geschichte Rußlands.* Second volume. Zurich 1945

Gmelin, Johann Georg: *Flora sibirica sive historia plantarum Sibiriae*. 4 vols. St. Petersburg 1747–1769; German revision: *Reise durch Sibirien von dem Jahr 1733 bis 1743*. 4 vols. Göttingen 1751–1752, cf. also Posselt

Goldstine, Hermann H.: *A History of the Calculus of Variations from the 17th through the 19th Century*. New York, Heidelberg, Berlin 1980 (Studies in the History of Mathematics and Physical Sciences 5)

Goethe, Johann Wolfgang von: *Materialien zur Geschichte der Farbenlehre. Achtzehntes Jahrhundert. Zweite Epoche*. Article *Achromasie*. In: Gedenkausgabe der Werke, Briefe und Gespräche. Edited by E. Beutler. Vol. 16. Zurich 1949

Grau, Conrad: *Berühmte Wissenschaftsakademien*. Leipzig 1988

Groehler, Olaf: *Die Kriege Friedrichs II*. Berlin 1989

Habicht, Walter: *Einleitung zu Band 20 der zweiten Serie*. In: O.II,20, p. VII–LX (1974)

–: *Leonhard Eulers Schiffstheorie*. In: O.II,21, p. VII–CCXLII (1978)

–: *Einige grundlegende Themen in Leonhard Eulers Schiffstheorie*. In: EGB 83, p. 243–270 [1983a]

–: *Betrachtungen zu Eulers Dioptrik*. In: EGB 83, p. 283–302 [1983b]

Hamel, Georg: *Elementare Mechanik*. Leipzig und Berlin 1912

Harnack, Adolf: *Geschichte der Königlich-Preußischen Akademie der Wissenschaften zu Berlin*. 3 vols. in 4. Berlin 1900

Hartweg, F. G.: *Leonhard Eulers Tätigkeit in der französisch-reformierten Kirche von Berlin*. In: Die Hugenottenkirche 32 (1979), Nr. 4, p. 14f., Nr. 5, p. 17f.

Helmholtz, Hermann von: *Die Lehre von den Tonempfindungen*. 6th ed. Braunschweig 1913

Hermann, Jacob: *Phoronomia, sive de viribus et motibus corporum solidorum et fluidorum libri duo*. Amsterdam 1716

Herzog, J. W.: *Athenae Rauricae*. Basel 1778 (Appendix 1780)

Hildebrandt, Stefan: *Euler und die Variationsrechnung*. In: Zum Werk Leonhard Eulers. Vorträge des Euler-Kolloquiums im Mai 1983 in Berlin. Edited by E. Knobloch et al. Basel, Boston, Stuttgart 1984, p. 21–35

IMI: The following volumes of this journal contain papers about Euler: II, V–VII, X, XII, XIII, XVI–XXX, XXXII/XXXIII

Jänisch, Carl Friedrich von: *Traité des applications de l'analyse mathématique au jeu des échecs*. 3 vols. St. Petersburg 1862/63

Jaquel, Roger: *Le savant et philosophe Mulhousien Jean-Henri Lambert (1728–1777)*. Études critiques et documentaires. Paris 1977

–: *Leonhard Euler, son fils Johann Albrecht et leur ami Jean III Bernoulli jusqu'en 1766*. In: EGB 83, p. 435–446

Kaiser, Wolfram: *Johann Andreas Segner*. Leipzig 1977 (Biographien hervorragender Naturwissenschaftler, Vol. 31)

Kneser, Adolf: *Euler und die Variationsrechnung*. In: Festschrift 1907 (s. d.), p. 21–60

Knobloch, Eberhard: *Musiktheorie in Eulers Notizbüchern*. In: NTM-Schriftenreihe für Geschichte der Naturwissenschaften, Technik und Medizin, 24 (1987), Fasc. 2, p. 63–76

Knobloch, Wolfgang: *Leonhard Eulers Wirken an der Berliner Akademie der Wissenschaften 1741–1766*. Berlin 1984 (Studien zur Geschichte der Wissenschaften der DDR, Vol. 11)

König, Johann Samuel (Translator): *Figur der Erden bestimmt durch die Beobachtungen des Herrn von Maupertuis.* . . . Zurich 1741, 2d ed. 1761

Kopelevich, Yudif Ch.: *Pis'ma pervykh akademikov iz Peterburga*. Vestnik AN SSSR, Nr. 10 (1973), p. 121–133

–: *Vozniknovenie nauchnikh akademiy seredina XVII–seredina XVIII v Leningrad 1974* (Russian)

–: *Osnovaniye Peterburgskoy Akademiy nauk*. Leningrad 1977 (Russian)

Kovalevskaya, Sophia W.: *Sur le problème de la rotation d'un corps solide autour d'un point fixe*. In: Acta mathematica 12 (1889), p. 177–232

Lagrange, Joseph-Louis: *Mécanique analytique*. In: Œuvres, Vol. 12. Paris 1889 (first edition 1788)

Laugwitz, Detlef: *Infinitesimalkalkül. Kontinuum und Zahlen. Eine elementare Einführung in die Nichtstandard-Analysis*. Mannheim, Vienna, Zurich 1978

–: *Die Nichtstandard-Analysis: Eine Wiederaufnahme der Ideen und Methoden von Leibniz und Euler*. In: EGB 83, p. 185–198

–: *Zahlen und Kontinuum. Eine Einführung in die Infinitesimalmathematik*. Mannheim, Vienna, Zurich 1986 (Lehrbücher und Monographien zur Didaktik der Mathematik, Vol. 5)

Maclaurin, Colin: *Treatise of Fluxions*. Edinburgh 1742

Marpurg, Friedrich Wilhelm: *Versuch über die musikalische Temperatur*. Wrocław 1776

Mattheson, Johann: *Große Generalbaß-Schule*. Hamburg 1731

Maupertuis, Pierre-Louis de: *Figure de la terre déterminé par les observations de MM. de Maupertuis, Clairaut* [etc.]. Paris 1738 (German translation cf. König, J. S.)

Mikhailov, Gleb K.: *Leonard Eyler*. In: Izvestiya akademiy nauk SSSR. Otdeleniye tekhnicheskikh nauk, Nr. 1 (1955), p. 3–26 (Russian)

–: *K 250–letiyu so dnya rozhdeniya Leonarda Eylera*. In: Izvestiya Akademiy nauk SSSR. Otdeleniye tekhnicheskikh nauk, Nr. 3 (1957), p. 3–48 (Russian)

–: *Notizen über die unveröffentlichten Manuskripte von Leonhard Euler*. In: Sammelband Schröder (1959), p. 256–280

–: *Euler und die Entwicklung der Mechanik*. In: Abhandlungen der Akademie der Wissenschaften der DDR – Abteilung Mathematik, Naturwissenschaften, Technik. Jg. 1985, Nr. 1 N, Festakt. . . Leonhard Euler. Berlin 1985

–: *Leonhard Euler und die Entwicklung der theoretischen Hydraulik im zweiten Viertel des 18. Jahrhunderts.* In: EGB 83, p. 229–241

Mittenzwei, Ingrid: *Friedrich II. von Preußen.* Berlin 1984

Nagel, Fritz: *A Catalog of the Works of Jacob Hermann.* In: Historia Mathematica 18 (1991), p. 36–54

Nevskaya, Nina I.: *Leonhard Euler und die Astronomie.* In: EGB 83, p. 363–371

Newton, Isaac: *Philosophiae naturalis principia mathematica.* London 1687, 2d ed. 1713, 3d ed. 1726 (German: *Mathematische Prinzipien der Naturlehre.* With remarks and explanations edited by J. Ph. Wolfers. Berlin 1872, reprint Darmstadt 1963); *Die mathematischen Prinzipien der Physik,* translated and edited by Volkmar Schüller. Berlin, New York 1999

Ostrovityanov, K. W. (Ed.): *Istoriya Akademiy nauk.* Moscow, Leningrad 1958, 1964, 2 vols. (Russian)

Pekarski, P. P.: *Ekaterina II i Eyler.* Zapiski imperatorskoy akademiy nauk. 6th vol. Petersburg 1865 (Russian)

–: *Istoriya imperatorskoy akademiy nauk v Peterburge.* 2 vols. Petersburg 1870–1873 (Russian)

Petrov, A. N.: *Pamyatnye Eylerovskiye mesta v Leningrad.* In: Ceremonial volume Moscow 1958, p. 597–604 (Russian with German summary)

Philidor, François André Danican: *L'Analyse des Echecs.* London 1749

Posselt, Doris (Ed.): *Die Große Nordische Expedition von 1733 bis 1743. Aus Berichten der Forschungsreisenden Johann Georg Gmelin und Georg Wilhelm Steller.* Munich 1990

Pulte, Helmut: *Das Prinzip der kleinsten Wirkung und die Kraftkonzeptionen der rationalen Mechanik.* Stuttgart 1989 (Studia Leibnitiana, Sonderheft 19)

Raith, Michael: *Leonhard Euler—Ein Riehener?* In: Riehener Zeitung, December 21, 1979, Nr. 51/52, p. 9

–: *Gemeindekunde Riehen.* Edited by the Community Council Riehen. Riehen 1980

–: *Der Vater Paulus Euler—Beiträge zum Verständnis der geistigen Herkunft Leonhard Eulers.* In: EGB 83, p. 459–470

Reich, Karin: *Mathematik, Naturwissenschaften und Technik in Reclams Universal-Bibliothek 1883–1945.* In: Reclam—125 Jahre Universal-Bibliothek 1867–1992. Verlags- und kulturgeschichtliche Aufsätze. Edited by D. Bode. Stuttgart 1992, p. 148–166

Rimscha, Hans von: *Geschichte Rußlands.* Second revised and enlarged edition. Darmstadt 1970

Riehl, Alois: *Der philosophische Kritizismus.* Vol. 1, 2d ed. Leipzig 1908

Robins, Benjamin: *Remarks on Mr. Euler's Treatise of Motion...* London 1739

–: *New Principles of Gunnery.* London 1742

Schafheitlin, Paul: *Eine bisher ungedruckte Jugendarbeit von Leonhard Euler.* In: Sitzungs-berichte der Berliner Mathematischen Gesellschaft 21 (1922), p. 40–44

–: *Eine bisher unbekannte Rede von Leonhard Euler*. In: Sitzungsberichte der Berliner Mathematischen Gesellschaft 24 (1925), p. 10–13

Schmieden, Curt, Laugwitz, Detlef: *Eine Erweiterung der Infinitesimalrechnung*. In: Math. Zeitschr. 69 (1958), p. 1–39

Speiser, Andreas: *Leonhard Euler und die deutsche Philosophie*. Zurich 1934

Speiser, David: *Eulers Schriften zur Optik, zur Elektrizität und zum Magnetismus*. In: EGB 83, p. 215–228

Stäckel, Paul: *Johann Albrecht Euler*. In: Vierteljahresschrift der Naturforschenden Gesellschaft in Zürich. 55. Jg. 1910, Zurich 1910, p. 63–90

Staehelin, Andreas: *Geschichte der Universität Basel 1632–1818*. (Studien zur Geschichte der Wissenschaft in Basel. Published on the 500th anniversary of the University of Basel 1460–1960. IV/V, two parts). Basel 1957

– (Ed.): *Professoren der Universität Basel aus fünf Jahrhunderten*. Basel 1960

Stieda, Wilhelm: *Die Übersiedlung Leonhard Eulers von Berlin nach St. Petersburg*. In: Berichte über die Verhandlungen der Sächsischen Akademie der Wissenschaften zu Leipzig. Philologisch-historische Klasse, Vol. 83, 1931, Fasc. 3. Leipzig 1931, p. 4–62

–: *Johann Albrecht Euler in seinen Briefen 1766–1790*. In: Berichte über die Verhandlungen der Sächsischen Akademie der Wissenschaften zu Leipzig. Philologisch-historische Klasse, Vol. 84, 1932, Fasc. 1. Leipzig 1932, p. 4–43

Szabó, István: *Einführung in die Technische Mechanik*. Berlin 1975

–: *Geschichte der mechanischen Prinzipien und ihrer wichtigsten Anwendungen*. Basel 1987, Third corrected and enlarged edition, edited by P. Zimmermann and E. A. Fellmann

Timerding, Heinrich Emil: *Kant und Euler*. In: Kantstudien 23 (1919), p. 18–64

Truesdell, Clifford: *Rational Fluid Mechanics 1687–1765*. In: O.II,12 (1954)

–: I. *The First Three Sections of Euler's Treatise on Fluid Mechanics (1766)*; II. *The Theory of Aerial Sound, 1687–1788*; III. *Rational Fluid Mechanics, 1765–1788*. In: O.II,13 (1956)

–: *Essays in the History of Mechanics*. Berlin 1968

Van der Linde, Antonius: *Geschichte und Literatur des Schachspiels*. 2 vols. Berlin 1874

Vavilov, Sergey I. (Ed.): *Polnoye Sobraniye Sochineniy*. 10 vols. Moscow, Leningrad 1950–1959

Volk, Otto: *Eulers Beiträge zur Theorie der Bewegungen der Himmelskörper*. In: EGB 83, p. 345–362

Werckmeister, Andreas: *Musicalische Temperatur*, 2d ed. Frankfurt a. M., Leipzig 1691

Winter, Eduard (Ed.): *Die Registres der Berliner Akademie der Wissenschaften 1746–1766*. Berlin 1957 [*Registres*]

–: *Die deutsch-russische Begegnung und Leonhard Euler*. Berlin 1958

Wolfers, Jacob Ph. (Translator): *Leonhard Euler's Mechanik oder analytische Darstellung der Wissenschaften von der Bewegung....* Greifswald 1848–1850

Yushkevich, A. P.: *The Concept of Function up to the Middle of the 19th Century.* In: AHES 16 (1976/77), p. 37–85

– (Ed.): *Zur Theorie komplexer Funktionen—Arbeiten von Leonhard Euler....* Leipzig 1983 (Oswalds Klassiker der exakten Wissenschaften, Vol. 261)

–, Kopelevich, Yu. Ch.: *Christian Goldbach.* Basel 1994

# Testimonials

*Mikhail Vassilyevich Ostrogradski*
All famous mathematicians alive today, says a geometer who is as outstanding as he
is deep [LAPLACE], are students of EULER. There is not one among them who had not
benefitted by studying his works, who had not received from him the formulae and
methods needed, who in his discoveries was not guided and supported by his genius...
His formulae are simple and elegant; the clarity of his methods and proofs are further enhanced by a large number of well-selected examples. Neither NEWTON nor even
DESCARTES, as great as their influence has been, have achieved such fame as possessed,
among all geometers, by EULER alone, whole and undivided.

> From an expert opinion of April 2, 1844 on behalf of Minister UVALOV.
> German translation from the Russian by K.-R. BIERMANN in EGB 83, p. 492

*Niklaus Fuss*
His name, which history will place next to the ones of GALILEI, DESCARTES, LEIBNIZ,
NEWTON and so many other great men who have honored mankind with their genius,
can only expire with science itself.

> From the *Éloge*, 1783

*Georg Ferdinand Frobenius*
There is only one attribute of a perfect genius which EULER does not share: namely to
be unintelligible.

> Lecture in Basel, 1917

*Anton Friedrich Büsching*
LEONHARD EULER is not, like the great algebraists usually are, of sinister character and
clumsy behavior, but cheerful and lively (especially among acquaintances), and although his lost right eye looks somewhat disgusting, one soon gets used to it and finds
his face pleasant.

> Beyträge zur Lebensgeschichte denkwürdiger Personen.
> Part IV, Halle 1789

*Christian Wolff*
Mister EULER, who could enjoy his well-earned fame in higher mathematics, now wants
to dominate with a vengeance all the sciences, to which he was never predisposed, and
since he is lacking both the first principles and knowledge of the literature, which are
necessary for historical insight, he does great damage to his own fame, since there are
a few only who have an appreciation of the fame due to him.

> Letter of May 6, 1748 to Johann Daniel Schumacher

Testimonials

*Joseph Ehrenfried Hofmann*
EULER is one of the most astonishing personalities of the 18th century... Widely admired by some as the great teacher of Europe, who left his mark on the "mathematical century"; widely despised by others who want to see in him only a living computing machine and make fun of his peculiar philosophical views.

Physikalische Blätter 14 (1958)

*Eduard Fueter*
For where mathematical reason did not suffice, for EULER began the kingdom of God.

Die Geschichte der exakten Wissenschaften
in der Schweizer Aufklärung, 1941

*Arthur Schopenhauer*
Read only, e. g., in EULER's Letters to a Princess his exposition of the basic truths of mechanics and optics... if one reads them, it is as if one had exchanged a bad telescope against a good one.

Die Welt als Wille und Vorstellung. 2d Part, 1st Book, Chap. 15

*Georg Christoph Lichtenberg*
If from the great EULER's works one would take away all those things that have no immediate practical application, they would shrink tremendously. Yet, the great man has much occupied himself with the highest abstractions, which only future generations will know how to interpret.

Letter to Gottfried August Bürger, Fall (?) of 1787

*Herbert Weiz*
It is not rare that from problems posed to him, EULER received stimuli for deepening his mathematical work. Conversely, in his numerous theoretical investigations, he never lost sight of the practical aspects.

From the ceremonial address in Berlin (DDR) 1983

*Andreas Speiser*
If one considers the intellectual panorama open to EULER, and the continual success in his work, he must have been the happiest of all mortals, because nobody has ever experienced anything like it... Still not by far all that EULER had discovered has passed into the pool of common mathematical knowledge, and every time one goes fishing in his works, one has a rich catch. His views of mathematics have survived most of what was held in the 19th century, and is becoming more modern with each day.

Die Basler Mathematiker (1939)

# Index of names

Ackeret, Jakob, 45
Alembert, Jean le Rond d', 98, 106 f, 112, 122, 137
Anna Ivanovna, Tsarina 32, 56
Archimedes, 86
Argens, Jean Baptiste de Boyer, 75
August Wilhelm, brother of Frederick II, 92

Baudan, Charles de 63
Bauer, General, 126
Baumgarten, Alexander Gottlieb, 75
Bayer, Gottlieb Siegfried, 34
Bell, Carl von, 125
Benedetti, Giovanni 67
Bering, Vitus, 34
Berkeley, George, 74
Bernoulli, Daniel, 3 f, 6, 14, 17, 26 f, 31, 35, 39, 59, 67, 75, 99, 107
Bernoulli, Daniel II, 17
Bernoulli, Jakob I, xiii, 1, 5, 10, 12, 17, 20, 43, 65
Bernoulli, Jakob II, 17
Bernoulli, Johann I, xiii, 2, 5, 13 f, 17, 20, 27 f, 32, 50 f, 59 f, 65, 67,ff, 84, 117
Bernoulli, Johann II, 17, 59 f, 83, 89 ff
Bernoulli, Johann III, 17, 117, 119 f
Bernoulli, Niklaus, "the older", 17
Bernoulli, Niklaus, "the painter", 17
Bernoulli, Niklaus I, 10, 17
Bernoulli, Niklaus II, 3, 6, 17, 26 f, 33
Bernoulli, Niklaus t.y., 17
Bernoulli, René, 40
Biot, Jean-Baptiste, 96
Blumentrost, Laurentius, 27, 36
Bolzano, Beernard, 138
Borcke, Eleanor von, 87
Borcke, Friedrich Wilhelm von, 87

Bosshart, Beatrice, 54 f
Bouguer, Pierre, 20, 85
Brand, Bernhard, 9
Bratteler, Rosa, 132
Brevern, Karl von, 57
Brucker, Johann Georg, 40
Brucker, Johann Heinrich, 9
Brucker, Margaretha – *see* Euler, Margaretha
Brucker-Faber, Maria Magdalena, 1, 14
Bülfinger, Georg Bernhard, 4, 31, 33, 36 f
Burckhardt, Johannes, 14
Burya, Abel, 131
Busch, Hermann Richard, 48
Buxtorf, Johannes, 9

Camus, Charles-Étienne-Louis, 20, 85
Cantor, Georg, 138
Cassini, Gian Domenico 85
Castillon, Frédéric, 110
Castillon, Jean (Gian Francesco Salvermini), 111
Catherine I, Tsarina, 3, 26 f, 31
Catherine II, Tsarina, 106, 115, 116, 119, 130
Catt, Henri Alexandre de, 112
Cauchy, Augustin Louis, 45, 138
Celsius, Anders 85
Châtelet, Émile du, 81
Chernishev, Count, 102
Clairaut, 85, 89, 98, 104, 122, 137
Clairaut, Alexis-Claude, 84, 123
Columbus, Christoph, 86
Condorcet, M.-J.-A., Nicolas Caritat de, 131
Curione, Celio Secondo, 9

Dashkova, Yekaterina Romanovna, 130
Delen, Jakob van, 104, 111, 125

# About the author

Emil Alfred Fellmann, b. 1927 in Basel. Study of Mathematics, Astronomy, Theoretical Physics, and Philosophy at the University of Basel. Since 1954 specialization in the history of the so-called exact sciences with emphasis on the 17th and 18th century. Election in 1971 to *Membre Correspondant*, in 1983 to *Membre Effectif* of the *Académie Internationale d'Histoire des Sciences, Paris*. From 1972 to 1997 active member of the *Euler-Kommission der Schweizerischen Akademie der Naturwissenschaften* and from 1986 to 2006 editor-in-chief of the 4th Series of Leonhard Euler's *Opera omnia*, at the same time President of the *International Editorial Committee Switzerland-Russia*. Visiting Professor and lecturing activities at several European universities and international symposia. (Two honorary degrees: London 1959; Basel 2001).

*Publications*: Next to numerous publications in various technical periodicals and books, the monograph G. W. Leibniz – *Marginalia in* Newtoni *'Principia mathematica' (1687)*, Paris 1973; responsible editor and coauthor of the memorial volume Leonhard Euler *1707–1783. Beiträge zu Leben und Werk*, Basel 1983; coeditor of the volumes III 9 (1973) and IVA 2 (1998) of Euler's *Opera omnia*; editor of the book series *Vita Mathematica*, Birkhäuser Basel and coeditor of the latter's series *Science Networks · Historical Studies*. Contributions to several encyclopediae and dictionaries.

## Thanks

My heartfelt thanks go to all institutions and private persons named in the Sources for the illustrations for the generously granted reproduction rights. The lion's share of reproductions I owe to the manager of the department of reproductions at the University Library of Basel, Mr. Marcel Jenni, who in 1983 was also responsible for the graphical illustrations in the truly beautiful Basel memorial volume for Euler, to which he engaged himself indefatigably and unselfishly. Finally, I extend my especially cordial thanks to Prof. Gleb Konstantinovich Mikhailov (Moscow) for letting me use the text on p. 124–129, an original contribution of unquestionably special zest.

# Sources for the illustrations

Universitätsbibliothek Basel. Reprophotographie: ii, x, 8, 12, 13, 24, 26(4),
   29, 30, 44, 46 left, 66, 68, 69, 74, 81, 82, 83, 84, 91, 97, 107, 109, 110,
   119, 121, 123, 132
Universitätsbibliothek Basel. Handschriftenabteilung: 15, 28, 49, 108
Archiv für Kunst und Geschichte, Berlin: 16 left, 116
By courtesy of the National Portrait Gallery, London: 16 right
After: Jahrbuch Überblicke Mathematik 1993. Edited by S. D. Chatterji et
   al. Braunschweig, Wiesbaden: 17
Euler-Archiv Basel: 19, 46 right, 47, 52(3), 71, 88, 103, 117, 118
Bildarchiv Preußischer Kulturbesitz, Berlin: 25, 86
Archive of the Academy of Sciences of Russia, St. Petersburg: 38, 70
From: Georg Holmsten: Voltaire. Reinbek bei Hamburg 1971: 61
From: W. Ahrens: Mathematische Unterhaltungen und Spiele. Vol. I. Leip-
   zig, Berlin[3] 1921: 79(2) bottom
Musée d'Art et d'Histoire, Geneva: 127
Lomonosov-University Moscow: 130
Private archive E. A. Fellmann: (Photo: Christian Fellmann): 137(2)